01-WSM-3

RESEARCH DIGEST

BIOASSESSMENT: A TOOL FOR MANAGING AQUATIC LIFE USES FOR URBAN STREAMS

by:

Michael T. Barbour
Michael J. Paul
David W. Bressler
Center for Ecological Sciences
Tetra Tech, Inc.

Alison Purcell O'Dowd
Vincent H. Resh
University of California, Berkeley

Ed Rankin
Midwest Biodiversity Institute

2007

WERF

The Water Environment Research Foundation, a not-for-profit organization, funds and manages water quality research for its subscribers through a diverse public-private partnership between municipal utilities, corporations, academia, industry, and the federal government. WERF subscribers include municipal and regional water and wastewater utilities, industrial corporations, environmental engineering firms, and others that share a commitment to cost-effective water quality solutions. WERF is dedicated to advancing science and technology addressing water quality issues as they impact water resources, the atmosphere, the lands, and quality of life.

For more information, contact:
Water Environment Research Foundation
635 Slaters Lane, Suite 300
Alexandria, VA 22314-1177
Tel: (703) 684-2470
Fax: (703) 299-0742
www.werf.org
werf@werf.org

This report was co-published by the following organizations. For non-subscriber sales information, contact:

IWA Publishing
Alliance House, 12 Caxton Street
London SW1H 0QS, United Kingdom
Tel: +44 (0) 20 7654 5500
Fax: +44 (0) 20 7654 5555
www.iwapublishing.com
publications@iwap.co.uk

Library of Congress Catalog Card Number: 2006926766
IWAP ISBN : 978-1-84339-788-5

Tetra Tech, Inc.

The research on which this report is based was developed, in part, by the United States Environmental Protection Agency (EPA) through Cooperative Agreement No. CR831559-01with the Water Environment Research Foundation (WERF). However, the views expressed in this document are solely those of Tetra Tech, Inc. and neither EPA nor WERF endorses any products or commercial services mentioned in this publication. This report is a publication of WERF, not EPA. Funds awarded under the Cooperative Agreement cited above were not used for editorial services, reproduction, printing, or distribution.

ACKNOWLEDGMENTS

The project team acknowledges numerous individuals who provided assistance in various areas of the project. Drs. Jeroen Gerritsen, Lei Zheng, and Sam Stribling of Tetra Tech provided critical direction on design, analyses, and review. Sabu Paul and Dr. Ting Dai, also of Tetra Tech, were the primary designers of the Flow Time Series Estimation tool used to generate hydrologic data for ungauged stream locations (which was submitted as a separate product under this research contract). We would also thank the Project Subcommittee for their comments and insights. Special thanks go to Margaret Stewart for her guidance through the duration of the 3-year project.

The project team also acknowledges and thanks individuals from each of the three study regions.

Baltimore

From the Baltimore area, we thank Dr. Ron Klauda and Marty Hurd at Maryland Department of Natural Resources; Bill Stack and Ted Eucare at Baltimore City Department of Public Works; Steve Stewart at Baltimore County Department of Environmental Programs; Angela Morales at Howard County Department of Public Works Storm Water Management Division; Sharon Meigs at Prince George's County Department of Environmental Resources; Keith Van Ness at Montgomery County Department of Environmental Protection; and Chris Victoria at Anne Arundel County Office of Environmental and Cultural Resources. These individuals provided water chemistry, physical habitat, and biological data.

San Jose

From the San Jose study region, we thank Dr. James Carter and Steven Fend of the United States Geological Survey for the use of their biological data from the Santa Clara Basin (San Jose, California). We particularly thank James Carter for his insight and guidance in refining the methods for the selection of biological indicators. We also thank Dr. Faith Kearns (University of California, Berkeley) for the use of her habitat assessment and water quality data. Last, we thank Paul Randall and Chris Sommers (EOA, Inc.) for their help in acquiring existing data from the Santa Clara Basin.

Cleveland

From the Cleveland study region, we thank the numerous individuals from the Ohio Environmental Protection Agency who provided data, especially Dennis Mishne and Jeff DeShon and the biologists at the Ecological Assessment Unit who provided guidance and insight on interpreting urban impacts in the Cleveland area. Bob Miltner of Ohio Environmental Protection Agency performed some early analyses on the Ohio data. The Institute for Local Government Administration and Rural Development at Ohio University in Athens, provided the geographic information system data for sites in the Cleveland area.

Research Team:

Principal Investigator:
Michael T. Barbour, Ph.D., *Center for Ecological Sciences, Tetra Tech, Inc.*

Project Team:
Michael J. Paul, Ph.D., *Center for Ecological Sciences, Tetra Tech, Inc.*
David W. Bressler, *Center for Ecological Sciences, Tetra Tech, Inc.*
Ed Rankin, *Midwest Biodiversity Institute*
Alison Purcell O'Dowd, *University of California, Berkeley*
Vincent H. Resh, Ph.D., *University of California, Berkeley*

Project Subcommittee:
Michael Bastian (Chair), *CH2M Hill*
Scott Dyer, Ph.D., *The Procter & Gamble Company*
Irwin Polls, formerly of *Metropolitan Water Reclamation District of Chicago*
Robin Reash, *American Electrical Power Corporation*
Paul Stacey, *Connecticut Department of Environmental Protection*
William Swietlik, *United States Environmental Protection Agency, Office of Water*

Water Environmental Research Foundation Staff

Director of Research:	Daniel M. Woltering, Ph.D.
Program Director:	Margaret Stewart

ABSTRACT AND BENEFITS

Abstract:

This research addresses the utility of bioassessment for managing aquatic life uses in urban and/or urbanizing catchments (i.e., watersheds). Heavily urbanized catchments present a problem for facilities and water quality managers struggling to balance the socio-economic needs of urban areas with aquatic life use standards. Most standards either do not recognize the limitations on achievable biological condition in urban areas, or set attainment goals too low by not addressing biological potential for improvement. This research specifically defines a process for developing alternative biological benchmarks for aquatic life use in urban catchments.

This research was conducted across three distinct climatic regions and describes a three-step process: 1) developing a primary urbanization gradient, 2) assembling an appropriate urban biological index, and 3) defining a biological potential that describes the highest biological condition currently achieved along the urban gradient. The primary urban gradient was developed using simple landscape and socio-economic measures of urbanization. Alternative urban gradients, comparable to the primary gradient, are presented that can be used as data and resources are available. The primary biological index was developed using a subset of commonly measured biological metrics. Last, biological potential was defined using quantile regression to characterize the upper boundary on biological condition observed along the primary urban gradient. This approach establishes empirically defined and realistic aquatic life use benchmarks for urbanized catchments, and describes a process by which the aquatic life use status of waterbodies in urbanized catchments can be placed in a realistic context. Guidance on implementation is provided for WERF subscribers for use in their particular urban areas.

Benefits:

- Defines a nationally applicable measure of urbanization.
- Defines a process for developing urban-specific biological indicators.
- Establishes empirically defined and realistic aquatic life use benchmarks for urban areas.
- Provides dischargers with a tool to address the relative aquatic life use condition of receiving waters in urban areas.
- Provides municipalities and agencies with a tool to prioritize waterbodies for management action in urban catchments.
- Provides a process for identifying likely targets for management activity to improve ecological condition of aquatic life uses in urban areas.

Keywords: Urban gradient, reference, biological condition, biological potential, bioassessment, aquatic life use, urbanization, stressor gradient.

TABLE OF CONTENTS

LIST OF TABLES

LIST OF FIGURES

Executive Summary

The Issues Pertinent to Urban Watersheds

Urban catchments (i.e., watersheds) are invariably influenced by chemical input from discharges and runoff, as well as nonchemical stressors, such as land development along the stream corridors and within the catchment. It has proven difficult to isolate stressors and describe their influence on urban ecosystems. This, in turn, has made it challenging to design assessment approaches that lead to effective restoration and protection plans for urban areas. Urban ecosystem assessors, therefore, must develop measures of urbanization and appropriate response indicators that consistently describe the major disturbances in the urban landscape.

The Solution

Bioassessment—the evaluation of a waterbody using biological surveys and other direct measurements of the resident biota in surface waters—has long been considered the most appropriate means of evaluating the health of an ecosystem for supporting aquatic life. Heavily urbanized catchments present a problem for facilities and water quality managers struggling to balance the socio-economic needs of urban areas with aquatic life use standards. Most standards either do not recognize the limitations on achievable biological condition in urban areas, or set attainment goals too low by not addressing biological potential for improvement.

This research addresses the utility of bioassessment for managing aquatic life uses in urban and/or urbanizing catchments and specifically defines a process for developing viable biological benchmarks for aquatic life use in urban catchments. The research was conducted across three distinct climatic regions and describes a three-step process: 1) developing a primary and alternative urbanization gradients, using simple landscape and socio-economic measures of urbanization and other data and resources, as available; 2) assembling an appropriate urban biological index, using a subset of commonly measured biological metrics; and 3) defining a biological potential, using quantile regression, that describes the highest biological condition currently achieved along the urban gradient.

What this Research Digest Provides

- An assessment tool for characterizing stressor (Step 1) and biological (Step 2) gradients and defining biological potential (Step 3) of urban streams in a context relevant to aquatic life uses and that is applicable in different regions of the country.
- Guidance on how to develop and calibrate a scientifically valid biological potential model for restoration goals.
- Supplemental tools, such as a comprehensive scientific literature database and an innovative Flow Time Series Estimation (FTSE software package; Tetra Tech, Inc., 2006).

Benefits and Applications of the Bioassessment Tool

- Describes a nationally applicable measure of urbanization and its effect on aquatic systems.
- Outlines a process for developing urban-specific biological indicators.
- Establishes empirically defined and realistic aquatic life use benchmarks for urban watersheds.
- Provides dischargers with a tool to address the relative aquatic life condition of receiving waters in urban areas.
- Equips municipalities and agencies with a tool to prioritize waterbodies for management action in urban catchments.
- Defines a process for identifying likely targets for management decisions to improve ecological condition of aquatic life uses in urban areas.

Frequently Asked Questions for WERF Subscribers

Q: What are the uses of the Research Digest and the accompanying tools?
A: Users can develop project guidance for developing realistic criteria for aquatic life uses in urban watersheds. This guidance can be used by agencies to establish restoration goals.

Q: Of what benefit is this product for facilitating cooperation among watershed stakeholders?
A: This product can be used to demonstrate the merits of bioassessment and the need to collect necessary data. These data can then be applied using the methods described in the report to develop region-specific aquatic life use criteria relevant to urbanized watersheds.

Q: What are the benefits of bioassessment compared to chemical concentration-only based approaches for assessing water quality?
A: Bioassessment is capable of detecting multiple stressors (chemical and nonchemical) in a more holistic manner than chemistry measurements (which can be highly variable both spatially and temporally). The biological potential and outlier analysis methods presented in Section 2.3 of the Research Digest provide a stepping stone to identifying the individual stressors affecting urban systems.

Q: If the indicators (biological and urban) provided in this digest do not seem applicable to a subscriber's region, will the Research Digest still be useful?
A: Yes. The most important lesson learned is the process for characterizing a relevant urban gradient or developing realistic biological endpoints in urban settings rather than specific attributes. In many areas, a biological index has been calibrated and validated, and can be used in the process described here.

Q: What is the value of the literature database provided as a supplement to the Research Digest?
A: This database is a comprehensive review of the most relevant scientific literature regarding the use of biological indicators of water quality, particularly focused on urbanization.

Q: What benefit does the FTSE tool provide as a supplement to the Research Digest?
A: The FTSE provides a powerful tool for predicting hydrologic behavior for any ungauged stream site.

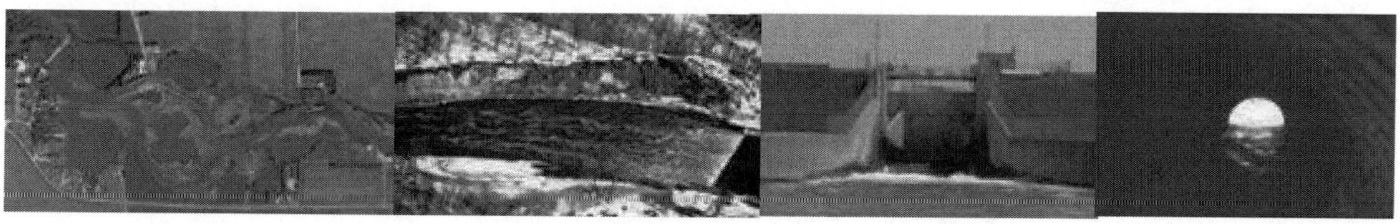

Stepwise Process for Development of Biological Potential Model

A three-step process is outlined below with brief descriptions of the relevant technical procedures and associated comments for consideration.

Stepwise Process	*Technical Procedures*	*Comments/Rationale*
Step 1 •Build *urban gradient of stressors*	•Calculate primary gradient = Average of: **a)** %urban LULC, **b)** road density, and **c)** population density by catchment.	•Alternative gradients – supplemental to primary, include chemistry and/or hydrology.

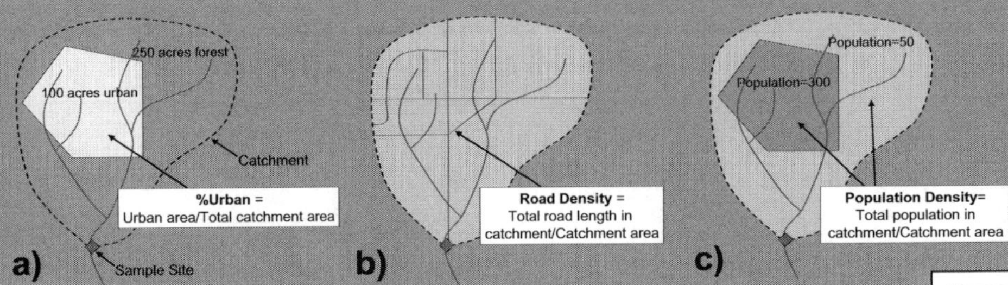

Research Digest § 2.1 pg. 2-1 to 2-7

Stepwise Process	*Technical Procedures*	*Comments/Rationale*
Step 2 •Develop urban ***biological index*** that responds to urban gradient	•Calculate individual metrics (EPT richness, Filterer richness, %Clinger), scale, and average to arrive at final biological index.	•The biological indicator developed here has been specifically designed for urban systems. In some regions, however, it is possible that an already existing state-developed index may be better suited for evaluating urban ecosystems.

- **EPT richness** = Total # of taxa in the classes Ephemeroptera (Mayflies), Plecoptera (Stoneflies), and Trichoptera (Caddisflies).
- **Filterer richness** – Total # of taxa in the filterer feeding group. These organisms feed by filtering fine particles of food from the water.
- **%Clinger** – Percentage of organisms in a sample that cling to rocks and other stream substrates. These organisms are relatively sensitive.

Research Digest § 2.2 pg. 2-7 to 2-13

Stepwise Process	*Technical Procedures*	*Comments/Rationale*
Step 3 •Define ***biological potential*** along urban gradient	•Define upper boundary of "wedge"-shaped relationship of biological indicator to primary gradient. •Outlier analysis to assess why some sites: ➢meet or exceed their potential ● ➢ while others are below potential ●	•This final analytical step defines the biological potential along the primary gradient of urban stress •Establishes a reasonable benchmark for all urban streams

Research Digest § 2.3 pg. 2-13 to 2-18

Steps 1-3

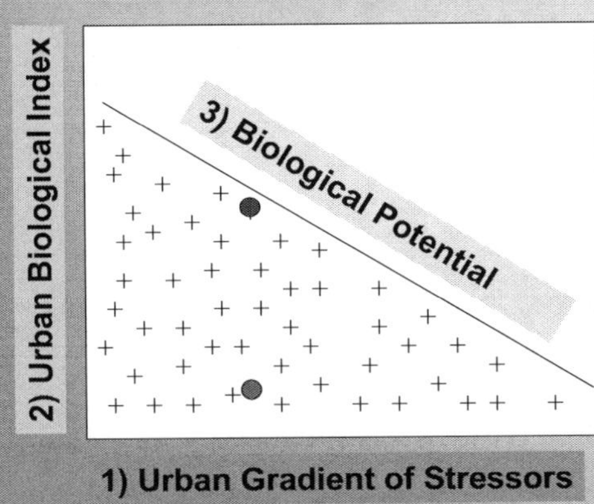

CHAPTER 1.0

PROJECT OVERVIEW

1.1 Project Background, Rationale, and Goals

The maintenance of healthy ecosystems in urban watersheds has long been a challenge for water quality managers because of multiple land use activities and cumulative stressors that occur in urban areas. Streams in urban settings are constantly influenced by the activities of densely populated catchments (Paul & Meyer, 2001). This relationship, recently termed the "urban stream syndrome" (Meyer et al., 2005), has confounded the ability of environmental managers to accurately characterize the potential of urban streams. There is a critical need to begin looking at all water quality and environmental impacts on a catchment (i.e., watershed) basis and to more effectively involve local catchment stakeholders in decisions relative to determining designated uses and implementing controls [U.S. Environmental Protection Agency (U.S. EPA), 1996b; Lindsey et al., 1997].

Urban catchments are invariably influenced by chemical input from discharges and runoff, as well as nonchemical stressors, such as land development along the stream corridors and within the catchment. The activities with the greatest impacts on aquatic life in urban catchments include the wholesale alteration of catchment hydrology, loss and degradation of riparian habitat, direct instream habitat degradation via channelization and culverting, excessive sedimentation resulting from historical and recent land disturbance activities and stream bank erosion (strongly linked to riparian encroachment), and contributions of excessive nutrients, oxygen-demanding wastes, and toxic chemical pollutants via urban runoff, point source discharges (both permitted and unpermitted), spills, and other releases (Yoder et al., 1999). Isolation of the wide variety of stressors and a description their influence on urban ecosystems has made it challenging to design assessment approaches that lead to effective restoration and protection plans for these areas. Urban ecosystem assessment, therefore, must rely on developing both appropriate measures of urbanization and appropriate response indicators, both of which can be used consistently to describe the major disturbances in the urban landscape.

Bioassessment (i.e., the evaluation of the condition of a waterbody using biological surveys and other direct measurements of the resident biota in surface waters) has long been considered the most appropriate means of evaluating the health of an ecosystem for supporting aquatic life (Davis, 1995; Barbour, 1997). Bioassessment is rapidly becoming important for detecting impairment from cumulative stressors (Barbour, 1997; Karr & Chu, 1999). However, its broad use in urban systems has been hampered by the inability to establish reasonable thresholds that represent the best compromise between the ideal of restoring and protecting ecological integrity and the engineering decisions made to increase the value added for human growth and sustenance. The challenges include defining an appropriate benchmark for expected biological condition under a range of urban intensities and diagnosing specific stressors or partitioning the effects of multiple stressors for management decisions regarding the protection and restoration of the aquatic resource in urbanized catchments.

U.S. EPA has issued technical and program implementation guidance for the development of bioassessment approaches and criteria intended to address the components of ecological integrity (i.e., chemical, physical, and biological integrity) (U.S. EPA, 1990, 1996a, 2000; Gerritsen, 1998). The approach of bioassessment and biocriteria (i.e., narrative or numeric expressions that describe the biological condition of aquatic communities inhabiting waters of a designated aquatic life use) as promoted by U.S. EPA and state water quality agencies is the comparison of biological condition of a waterbody or segment to a reference condition, which is an expected condition of that waterbody when minimally influenced by anthropogenic disturbance. The reference expectation is typically a regional composite, not a single site, which takes into account natural waterbody classes, and natural variability. Reference conditions are based on near-natural characteristics that determine the biological potential. Given that most urban sites have lost a significant portion of their natural characteristics, the question is what is an appropriate expectation for urban stream biological condition?

Federal water quality regulations (40 CFR Part 131.10[g]) specify six general criteria for setting designated uses below fishable/swimmable standards as follows three of which are applicable here: 1) imposition of the criteria for a higher use would result in widespread, adverse socio-economic impacts, 2) the criteria are not attainable because of natural background conditions, and 3) the criteria are not attainable because of irretrievable, anthropogenic impacts. The issue is to determine an appropriate aquatic life use for urban systems and whether tiered aquatic life uses (TALU; U.S. EPA, 2005) should be used by state agencies to accommodate significantly altered systems. U.S. EPA is currently addressing this issue through a national initiative and technical workgroup to establish guidelines for implementing decision criteria for aquatic life use along a biological condition and stressor gradient relationship (see inset). This concept provides a common framework from which to measure attainment and potential. The research discussed here builds upon this TALU framework specifically for urban ecosystems.

Tiered Aquatic Life Use (TALU)
Use of Biological Information to Better Define Designated Aquatic Life Uses in State and Tribal Water Quality Standards: Tiered Aquatic Life Uses, draft (U.S. EPA, 2005).

The Draft document provides information that will help States and Tribes assess and communicate current biological conditions and progress towards meeting aquatic life protection goals. The document describes how to define aquatic life uses as tiers of biological condition, from natural to severely-altered. A waterbody's highest attainable use (tier) with its expected biological attributes and appropriate water quality criteria are determined by the waterbody's environmental setting. This tiered approach can help protect high-quality waters, while developing scientifically-sound expectations for restoration of damaged or impaired waters. The document is a result of several years of effort by a workgroup largely comprised of State and Tribal members, and draws from State programs that currently use a tiered approach to their aquatic life use designations. *Excerpt from http://www.epa.gov/bioindicators/html/bcg.html.*

The research presented here addresses technical issues related to bioassessment and biocriteria for urban streams, including application in water quality standards – specifically, aquatic life uses. Three general project objectives were used in guiding the research presented here:

- Develop a method for assembling biological indicators that are effective in describing the ecological status of urban systems.
- Develop a method for defining the potential of aquatic life across an urban gradient of stress.
- Investigate this potential across climatic zones.

This project follows closely upon its predecessor projects (Barbour et al., 1999a; WERF Report No. 97-IRM-1, *An Assessment of the Status, Usefulness, and Resource Requirements for Implementation of Biocriteria Water Quality Standards* and Paulson et al., 2001; WERF Report No. 98-WSM-1, *Distinguishing the Relative Influence of Habitat and Water Quality on Aquatic Biota*) and addresses concerns raised by WERF subscribers relevant to reference condition and aquatic use attainment for urban systems (see also WERF Project 91-NPS-1, *A Comprehensive UAA Technical Reference and User's Guide*). The research presented here is of national scope and encompasses urban catchments that are representative of three geographical regions: East, Midwest, and West. An approach is developed here to synthesize the mosaic of urban land use effects (e.g., catchment development history, chemical stressors, and catchment scale influences) that influence the health of aquatic systems into a single urban gradient, to develop an appropriate biological indicator of urbanization, and to empirically define an achievable biological potential for urban systems.

The benefits of this project are that it: 1) defines a measure of urbanization that is nationally applicable; 2) defines a process for developing urban-specific biological indicators; 3) establishes empirically defined and realistic aquatic life use benchmarks for urban areas; 4) provides dischargers with a tool to address the relative aquatic life use condition of receiving waters in urban areas; 5) provides municipalities and agencies with a tool to prioritize waterbodies for management action in urban catchments; and 6) provides a process for identifying likely targets for management activity to improve ecological condition of aquatic life uses in urban areas.

This research produced three products: 1) this research digest, which articulates the scientific process; 2) a literature database in interactive format for future reference and expansion; and 3) a documented hydrologic modeling tool for characterizing stream flow characteristics in ungauged urban streams.

1.2 Research Process

A three-step process (Figure 1-1) was used to systematically address the project objectives:

- Develop an urban gradient of stressors characterizing urban intensity (Section 2.1)
- Develop an understandable, nationally applicable biological indicator of urbanization (Section 2.2)
- Define the biological potential along the urban gradient (Section 2.3)

Each step builds on the results generated in the previous step. The stressor gradient represents the standard by which the performance of biological indicators is measured and biological potential is defined. The gradient(s) encompass various common components of urbanization at the stream and landscape levels. The biological indicators represent characteristics of benthic macroinvertebrate assemblages that are responsive to the urban stressors portrayed by these gradients. The biological potential describes the range of biological indicator values possible along the gradients of urbanization and provides a benchmark for expectations and restoration goals.

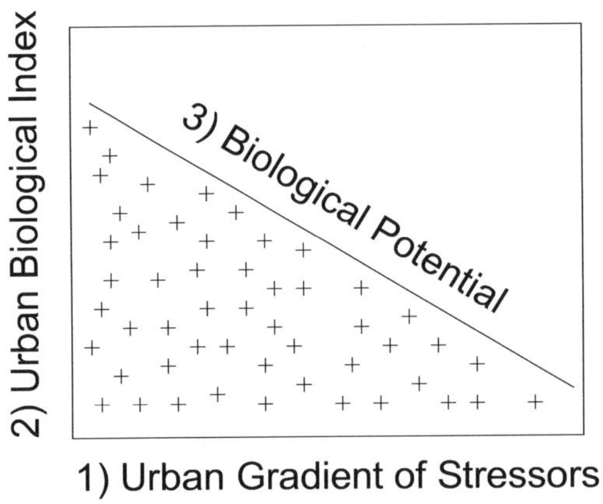

Figure 1-1. Model of Three Components Used for Addressing Project Goals (1) Stressor Gradient Along *x*-Axis; (2) a Biological Indicator Along the *y*-Axis; and (3) the Biological Potential Defined by the Relationship of the *x*- and *y*-Axes.

To address the broad-scale geographic applicability of the stressor gradients, biological indicators, and biological potential, three distinct regions (eastern, western, and midwestern United States) were examined (Figure 1-2). Analyses were first conducted on a comprehensive dataset (~3500 data points) from the Baltimore/Washington, DC, region of Maryland (East). These results were then validated using data from Cleveland, Ohio (Midwest) (~255 sites) and San Jose, Calif. (West) (~85 sites). To account for the influence of catchment size on biological indicators and biological potential data from a variety of stream sizes (1st–4th order) were used in the analysis. All data were collected from the mid 1990s to the present.

1.3 Data and Attachments

Landscape and socioeconomic data (variable list located in Appendix A) were summarized by catchment using a geographic information system. Chemistry, habitat, and biological data (Appendix A) were collected onsite by federal, state, county, and city agencies, as well as academic institutions. Raw biological data (benthic macroinvertebrate taxa identified to genus) were rarified to 100 organisms (to ensure consistency in sample size) and converted into metrics that described benthic assemblage composition (e.g., %Ephemeroptera), structure (e.g., %Clinger), and function (e.g., number of filterer taxa) (Appendix B) (Barbour et al., 1999b).

Hydrologic data were generated using a newly developed hydrologic modeling tool called Flow Time Series Estimation (FTSE, developed by Tetra Tech, Inc. and included on a CD along with user manual as an associated product with this Research Digest; Tetra Tech, 2006) that generates daily flow data for ungauged streams for a selected time (e.g., 20 years). There were only approximately 60 gauged sites available; therefore, the majority of sites used in the main analysis here were located on ungauged streams. The data produced by the FTSE tool were used to calculate hydrologic indicators with a downloadable Indicator of Hydrologic Alteration calculator that automatically generate a series of values that describe characteristics such as flow magnitude, duration, and flashiness (Richter et al., 1996; Baker et al., 2004). Because very few streams currently have hydrologic gauges and even fewer have consistent data over extended periods, this modeling tool proved to be highly useful as it allowed for the hydrology of a large number of catchments to be described and incorporated into the stressor gradients developed here.

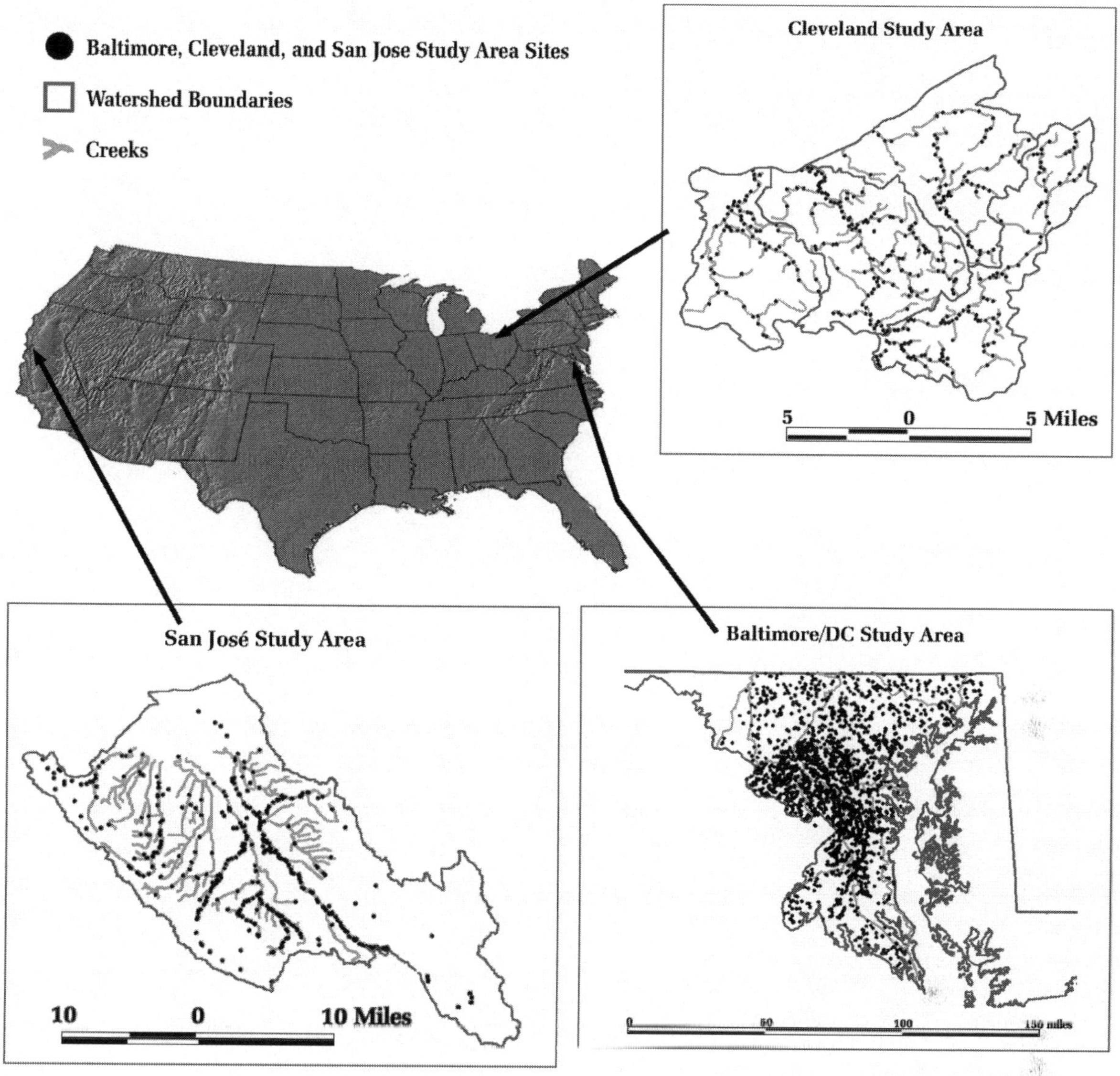

Figure 1-2. The Three Primary Study Regions and Sampling Sites: East (Baltimore/Washington, DC), West (San Jose, Calif.) and Midwest (Cleveland, Ohio) Study Regions.

A comprehensive literature database containing more than 700 citations was used in support of the analysis (available as a CD with user manual) and was consistently updated throughout the duration of the project. The literature database is in EndNote format to allow sorting by author, title, abstract, or keyword.

CHAPTER 2.0

Stepwise Process for Developing an Urban Biological Index and Defining Biological Potential

2.1 Step 1: Build an Urban Gradient of Stressors

Development of an appropriate gradient of urbanization requires integration of the diverse dimensions of urbanization into a synthetic measure. Several approaches have recently been used to represent urban gradients of disturbance. McMahon and Cuffney (2000) developed an urban index that used population density as the primary benchmark against which landscape and socioeconomic parameters were tested and selected for use in a final composite index. Population density was chosen as the benchmark because it explained the most variability in a principal components analysis of all available parameters. Tate et al. (2005) developed a similar urban intensity index in which landscape and socioeconomic variables were also selected for incorporation into the index based on their correlation with population density. Brown and Vivas (2005) developed a landscape intensity index that represents the energy used to establish and maintain different land uses. Numerous other studies have pointed to the use of impervious surface coverage as an effective indicator of urban intensity (Schueler, 1994).

An approach similar to those used by McMahon and Cuffney (2000) and Tate et al. (2005) is presented here. In this study three variables that represent sources of stress in urban systems (population density, road density, and percent urban land use/land cover) were standardized and averaged, and assembled into the primary urban gradient. These variables were selected based on the definition of urbanization and are intended to provide a more complete representation of urbanization than the benchmarks used in the aforementioned studies. Unlike McMahon and Cuffney (2000) who chose population density as the benchmark, the primary gradient presented here includes measures of urban infrastructure and industrial use, as well as population density.

Alternative urban gradients were then assembled using the primary urban gradient as the standard. These were constructed to provide stressor-based (instream), source-based (landscape), comprehensive (mixed stressor-source), cost-effective (built with easy to collect data), and desktop (built without field collected data) indicators that offer potential alternatives to the primary gradient that may be of interest to some users. The alternative gradients were built using landscape source and/or proximate (instream) stressor variables that were individually highly correlated with the primary gradient (Figure 2-1). These were then standardized and averaged in various combinations to form alternative urban gradients. Potential alternative gradients were then tested against the primary urban gradient. Those that were most highly correlated were selected as final alternatives (Figure 2-1).

The objective in this step was to develop primary and alternative gradients of urbanization that could be used for assessing biological response and ultimately for selecting a nationally applicable biological index (Step 2) and defining biological potential (Step 3). To achieve this objective, the gradients must represent the most prominent mosaic of stressors found in urban settings. These included instream stressor variables (instream characteristics such as water chemistry measures and hydrologic characteristics) and the landscape level sources of these stressors (including catchment and riparian urban land use/land cover (LULC), population density, and road density). Alternative gradients were developed and are presented here so that numerous options would be available for environmental managers according to the data types and resources that are available. In this way, managers can assemble the most robust and descriptive gradient within the limitations of the data available to them. The gradients were developed with the intent that they be applied throughout the country. For this reason, data from several regions (i.e., eastern, midwestern, and western United States) were used to calibrate and validate the urban gradient model.

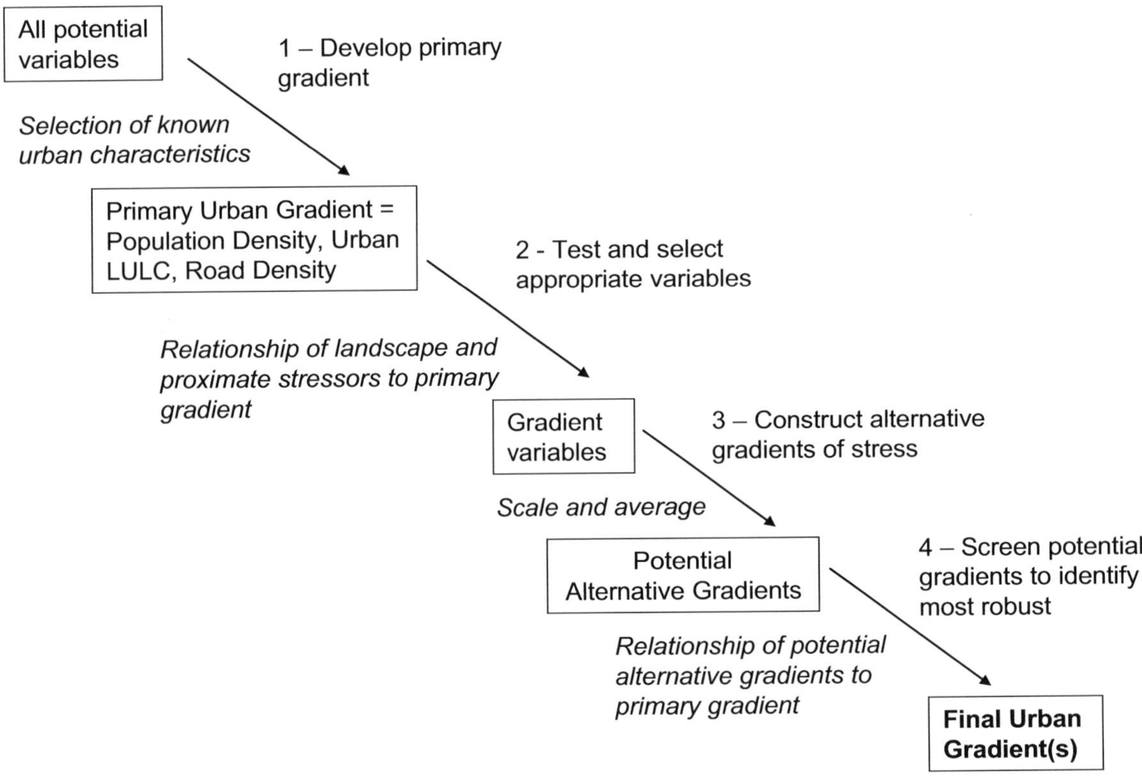

Figure 2-1. The Process for Building Primary and Alternative Urban Gradients. LULC = land use/land cover.

2.1.1 Develop a Primary Urban Gradient

A primary urban gradient was developed to represent the most important characteristics that typify an urban landscape. The primary gradient development process entailed defining urbanization and using first principles (direct measures of the very definition of urbanization) to choose quantitative measures that represented urbanization.

Urban: of, relating to, characteristic of, or constituting a city (Merriam-Webster Online: http://www.m-w.com/)

Urbanization: the process by which large numbers of people become permanently concentrated in relatively small areas, forming cities. (Encyclopedia Britannica Online: http://www.britannica.com/).

Elements associated with urbanization include concentrated populations, large settlements, and specialized labor. Three landscape-level parameters that characterize these definitions of urbanization were selected for the primary urban gradient:

- Population density (population/acre),
- Road density (m/acre), and
- Urban land use/land cover (LULC) (%)

These three parameters were estimated for each site catchment, standardized on a 100-point scale based on the 1st and 99th percentile for each variable, and these standardized scores averaged to produce the primary urban gradient.

2.1.2 Test and Select Appropriate Variables

The primary gradient was used as the principal urban gradient for developing an urban biological index (Step 2) and defining biological potential (Step 3). In addition, it was used to screen the instream stressors (chemistry, habitat, and hydrology) and landscape stressor sources (100 meter riparian LULC, demography, point sources) for use in the alternative urban gradients. Pearson product-moment correlations and multiple regressions were used to select parameters that were closely related to the primary gradient (i.e., indicative of urban stress, Appendix A). Using data from the Baltimore/Washington, DC, study area, these analyses resulted in the selection of five instream stressors and three landscape stressors (Table 2-1) for use in developing the alternative urban gradients.

Table 2-1. Instream and Landscape Stressors Selected for Incorporation into Alternative Urban Gradients.

Proximate Stressors		Landscape Variables
Chemistry	Hydrology	Riparian LULC (%)
Conductivity (µS/cm) Nitrate (mg/l)*	7-Day Minimum Flow High Pulse Duration Richard's-Baker Flashiness Index	Low Intensity Residential High Intensity Residential Commercial/Industrial

*Nitrate residuals after regression against agricultural land use (see Appendix A).

2.1.3 Construct Alternative Urban Gradients

The eight variables selected above were standardized on a 0–100 scale as for the primary gradient and averaged in different combinations to arrive at 17 different gradients (Table 2-2). The gradients were organized into five classes (Table 2-2):

1. *Stressor* – composed of instream stressors such as water chemistry parameters and hydrologic alterations (i.e., hydrologic indicators developed using FTSE)
2. *Source* – composed of landscape-level parameters that reflect the source of the stressors,
3. *Comprehensive* – composed of instream and landscape parameters,
4. *Cost-efficient* – composed of parameters that are easy to collect and/or assemble, and
5. *Desktop* – composed of parameters that can be computer generated.

Table 2-2. List of Alternative Urban Gradients and the Proximate and Landscape Stressors of Which These are Composed. Final candidate gradients selected are shaded.

		Proximate Stressors					Landscape Stressors		
		Chemistry		Hydrology			Riparian LULC (%)		
Gradient	Class	Conductivity (µS/cm)	Nitrate Residuals (mg/l)	7-Day Minimum	High Pulse Duration	R-B Flashiness	Low Intensity Residential	High Intensity Residential	Commercial/Industrial
Gradient 1	Stressor	x	x						
Gradient 2	Stressor	x	x		x	x			
Gradient 3	Stressor	x	x	x	x	x			
Gradient 4	Stressor				x	x			
Gradient 5	Stressor			x	x	x			
Gradient 6	Source							x	x
Gradient 7	Source						x	x	x
Gradient 8	Comprehensive	x	x		x	x		x	x
Gradient 9	Comprehensive	x	x	x	x	x	x	x	x
Gradient 10	Cost-efficient	x					x	x	x
Gradient 11	Cost-efficient	x			x	x	x	x	x
Gradient 12	Cost-efficient	x				x			
Gradient 13	Cost-efficient	x				x	x	x	x
Gradient 14	Cost-efficient	x			x	x			
Gradient 15	Desktop				x	x	x	x	x
Gradient 16	Desktop					x	x	x	x
Gradient 17	Comprehensive	x	x				x	x	x

2.1.4 Screen Potential Gradients

Correlation coefficients (*r*-values) (Appendix A), as well as evenness of the distribution, were used in selecting five final alternative gradients – one from each of the classes listed above (Table 2-2). Most of the gradients were similar to one another and showed similar relationships to the primary gradient (Figure 2-2); however, gradient 2, which is composed of only instream stressors (chemistry and hydrology), was noticeably less correlated with the primary urban gradient than the others.

2.1.5 Validate Gradient Models Across Geographic Regions

Western and midwestern U.S. data were used in assessing the broad-scale applicability of the primary and final alternative gradients. The goal in performing this validation was to assess the general applicability of urban gradients nationally. Nitrate was not collected in the West or Midwest; therefore, gradients 2 (stressor) and 9 (comprehensive) could not be compared regionally. The regional data (West and Midwest) were scaled using region-specific data ranges for the instream stressors (i.e., chemistry and hydrology), but scaled using eastern data ranges for the landscape parameters (e.g., Figures 2-3 and 2-4). The East study region encompassed the fullest range of values for the landscape parameters. Parameters that vary naturally among regions (i.e., instream stressors) were the only ones that needed to be scaled on a region-specific basis. Population density, road density, and urban LULC are entirely anthropogenic; therefore, there was no need to scale these parameters using region-specific ranges, as the scaling range reasonably approximated the potential range in landscape parameter levels (e.g., urban LULC 0–100%). In situations where instream stressor ranges are naturally different from another region it may be necessary to compensate for this variation using a correction factor. This coefficient would be based on the ranges of the instream stressors found in a particular geographic region and may be useful for more accurately portraying urban stress. After scaling, parameters were assembled into gradients as done previously. The gradient values for each region were plotted against the primary urban gradient values to confirm that all regional alternative gradients correlated similarly with the primary gradient. In fact, Figure 2-5 shows that the three gradients that could be compared were similarly correlated with the primary gradient among the three regions.

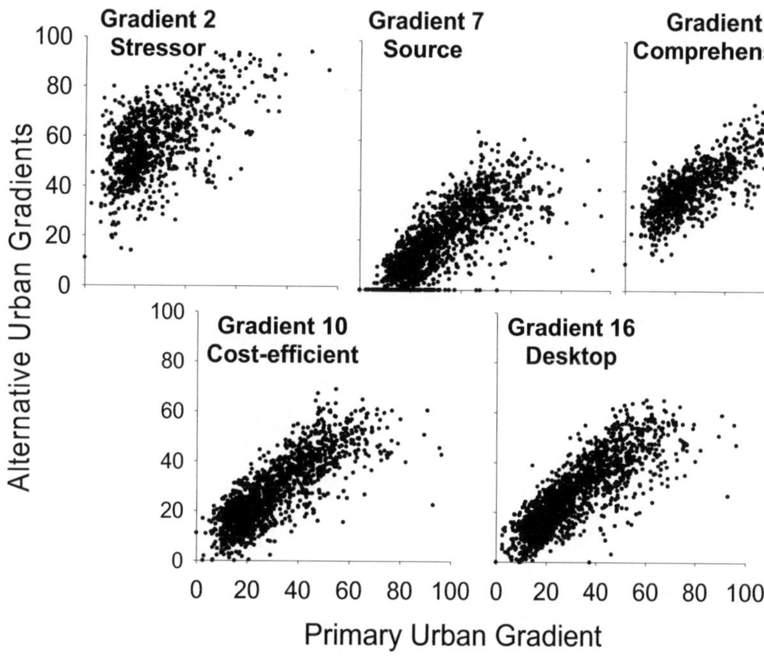

Figure 2-2. Five Final Alternative Stressor Gradients Plotted Against the Primary Urban Gradient.

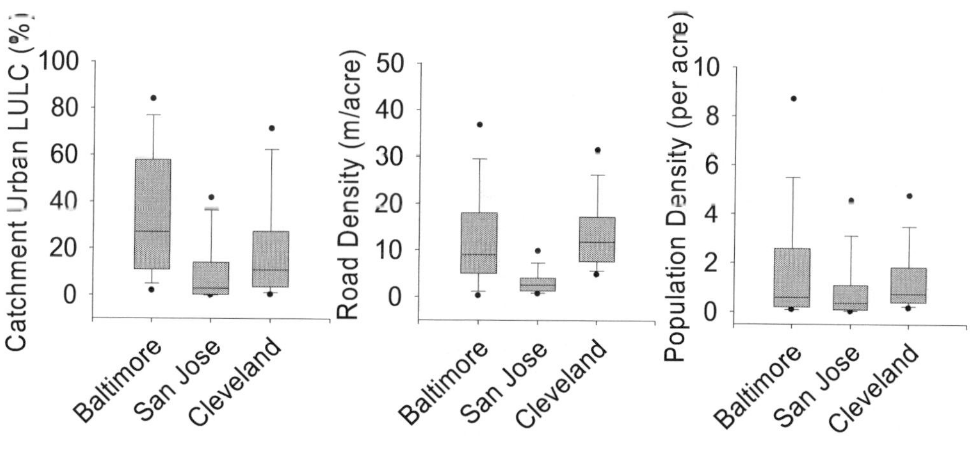

Figure 2-3. Ranges of Three Primary Urban Gradient Variables Among East, West, and Midwest Study Regions.

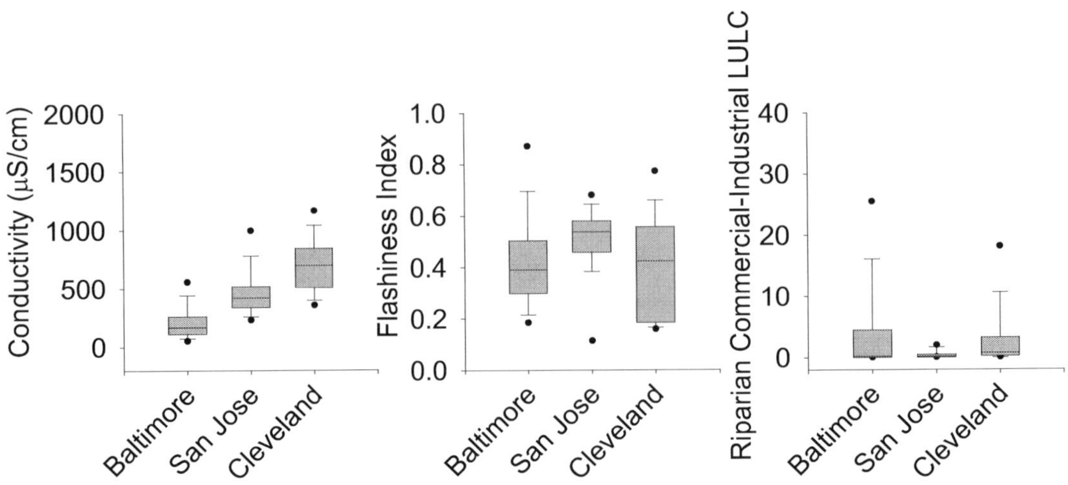

Figure 2-4. Ranges of Gradient Parameters for East, West, and Midwest Study Regions.

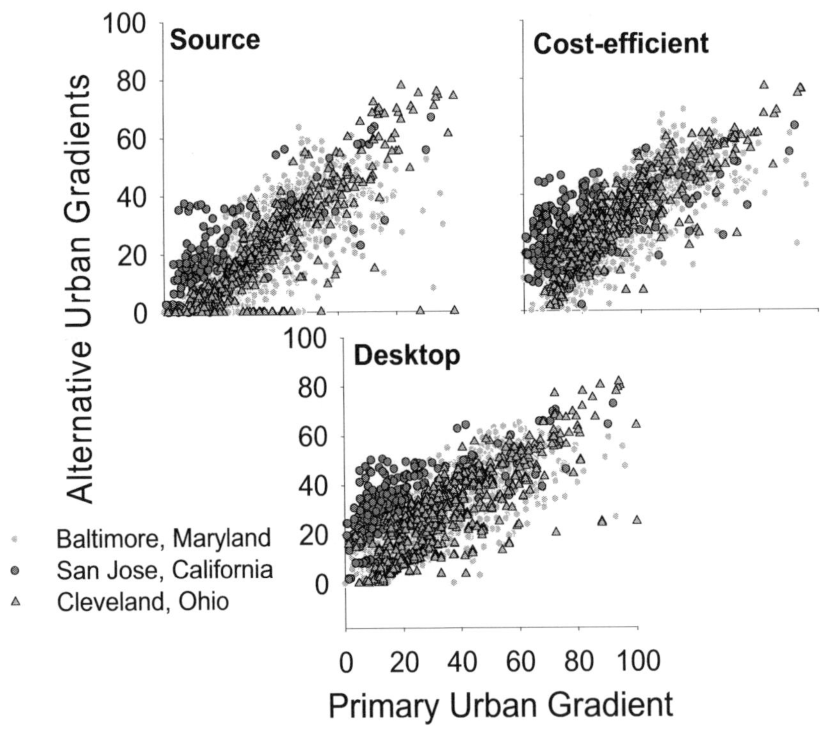

Figure 2-5. Three Alternative Gradients Plotted Against the Primary Urban Gradient for East, West, and Midwest Study Regions. All three alternative gradients shown here were positively and significantly ($P < 0.05$) correlated with the primary urban gradient. Source, cost-efficient, and desktop alternative gradient composition are described in Table 2-2.

2.1.6 Conclusions

In summary, the primary gradient was constructed from variables based on first principles regarding the effect of urbanization on landscapes. This primary gradient is used to develop the biological index (Step 2) and the biological potential (Step 3). The alternative gradients were developed to provide management options for use where data may be restricted or specific management goals require a different approach. All of the alternatives represent urbanization broadly (i.e., they respond to the primary gradient similarly across study regions). Ideally, gradients could be developed that contain all possible influential variables in urban systems and biological indicators could then be selected that reflect these multiple factors. However, because of limitations in time and resources, many agencies throughout the country measure only a subset of these parameters. Therefore, the gradients developed here are intended to incorporate a variety of parameters, some of which are commonly collected, and others that are difficult or expensive to collect and, thus, not often available for analysis. By developing the gradients in this way, they can be used in a wide variety of areas according to the data available in the specific area of interest.

2.2 Step 2: Develop an Urban Biological Index That Responds to Urban Gradient(s)

A variety of tools have been developed to serve as indicators of changes that occur in catchments receiving point and nonpoint discharges from domestic and industrial sources. Because detailed sampling of all potential biological, chemical, and physical features of a catchment on both a temporal and spatial scale is beyond the practical scope, ability, and budgets of agencies, indicators are used as a proxy to evaluate and characterize the quality and health of a stream system.

Of the aquatic assemblages having application to biological monitoring programs (fish, algae, macrophytes, invertebrates, bacteria, viruses, etc.), benthic macroinvertebrates are one of the most commonly used groups worldwide. As biological indicators, benthic macroinvertebrates can provide insight into the health of a waterbody, integrate the effects of cumulative stressors, and reflect current and past conditions (Barbour et al., 1999b). Furthermore, benthic macroinvertebrates are ubiquitous in most stream environments, have relatively long life histories and varying tolerances to perturbations in streams, and are cost-effective to use (Rosenberg & Resh, 1993; Yoder & Rankin, 1995). Several recent studies have validated the use of macroinvertebrate assemblages specifically as biological indicators of urban land use impacts (e.g., Walsh et al., 2005), demonstrating a consistent response to urbanization in regions across the globe.

This section describes the analyses used to characterize the response of benthic macroinvertebrate assemblages to the primary urban gradient (Section 2.1) through the development of a primary biological index (Figure 2-6). This indicator is then used as the standard by which the biological potential of streams along a gradient of urbanization is determined (Section 2.3).

These analyses began with a screening of biological metrics based on their relationships to the primary urban gradient. A metric is a characteristic of the biological assemblage that changes in a predictable way with human influence (e.g., taxa richness, percent tolerant individuals). The best performing metrics (i.e., those that are most responsive to the urban gradients) were assembled into potential indices based on three categories that describe the assemblage (community structure, ecological function, and habit). Responsive metrics were identified using linear response to the primary urban gradient. The urban gradient appeared to have its strongest effect on the maximum possible metric value rather than on the central tendency. This is likely caused by the variety of natural and nonnatural factors that affect biological condition at low urban levels (e.g., agriculture biotic interactions, biogeographic factors). Clearly, however, urbanization constrains biological condition. Quantile regression identifies the best fit through the selected quantile. The advantage of quantile regression is the capacity to identify relationships between the independent and dependent variables that may be different from that expressed by measures of the central tendency alone. In this example (Figure 2-7), the 95th quantile exhibits a steeper response to urbanization than the mean. Therefore, traditional linear regression was not a sufficient method to characterize the wedge-shaped relationship between the biological indices and the urban gradient (Figure 2-7). Instead, quantile regression (Koenker & Bassett, 1978) of the 95th quantile was used to characterize the upper boundary of the wedge-shaped scatterplots and to select the final biological index (Figure 2-8). This type of wedge-shaped plot is common in situations where there are many different variables influencing the *y*-axis and only a portion of these variables are represented by the *x*-axis. For instance, in this case, low biological index scores (i.e., low *y*-axis value) at sites with low urban intensity (i.e., low *x*-axis value) likely are influenced by other stressors such as agriculture that are not represented by the *x*-axis. Point source effects may also be a likely cause of depressed biology in catchments with otherwise minimal anthropogenic landscape disturbance.

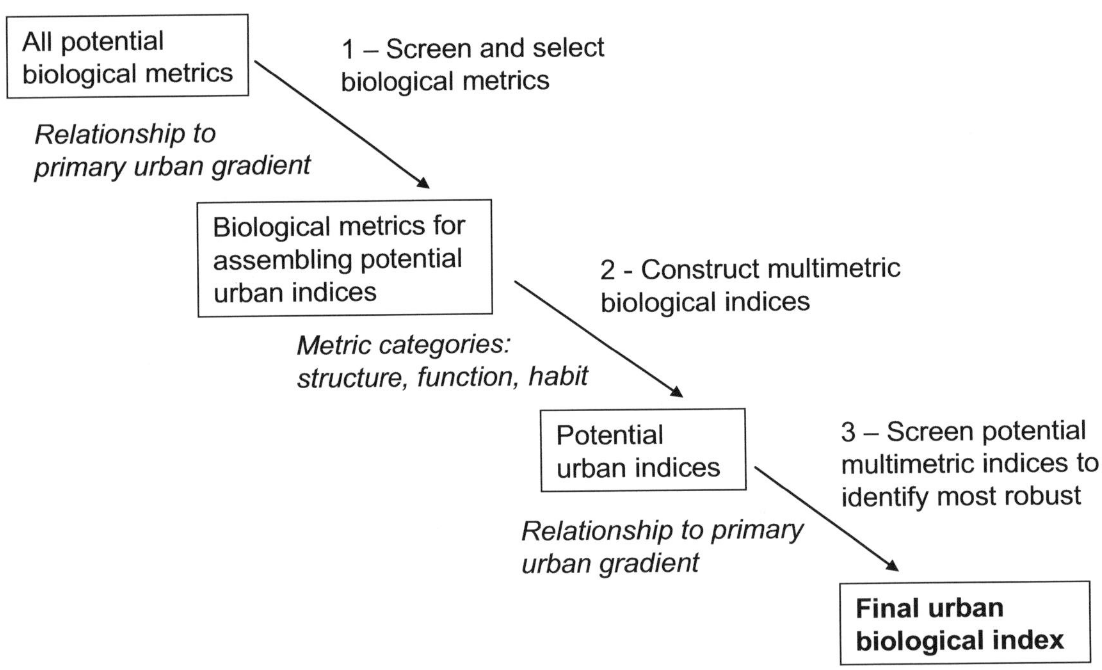

Figure 2-6. The Process for Developing Biological Index for Assessing Urban Streams Using Benthic Macroinvertebrate Data.

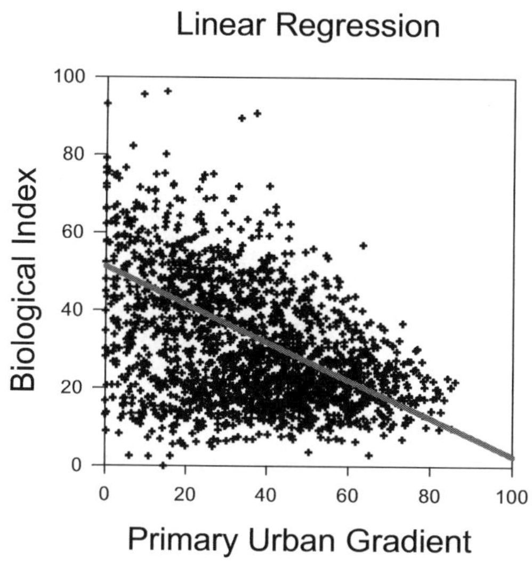

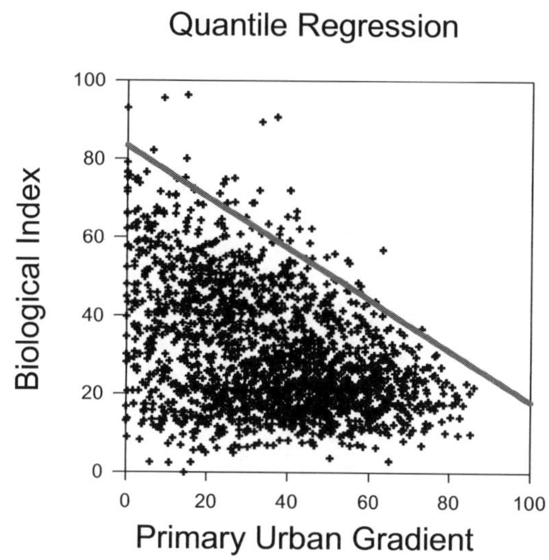

Figure 2-7. Example Scatterplots of a Biological Index (*y*-axis) Plotted Against Primary Urban Gradient (*x*-axis). The plot on the left shows a linear regression line (r^2= 0.19), while the plot on the right shows a 95% quantile regression line that better characterizes the upper boundary of the wedge-shaped plot.

2.2.1 Screen and Select Biological Metrics

Seventy two metrics were calculated from benthic macroinvertebrate data in each of the three study regions (Appendix B). By definition, these metrics should respond in a predictable manner to disturbance; in this case, urbanization. Seven types of metrics were calculated: 1) diversity (e.g., Shannon Diversity Index), 2) composition (e.g., percent Amphipoda), 3) richness (e.g., Ephemeroptera richness), 4) dominance (percent dominant taxa), 5) stressor tolerance (e.g., Hilsenhoff Biotic Index, percent tolerant), 6) functional feeding group (e.g., scraper, predator, filterer), and 7) habit (e.g., clinger, swimmer, climber) (Barbour et al., 1999b).

A four-step procedure was used to screen biological metrics for use in a biological index (Table 2-3).

Table 2-3. Summary of Criteria for Metric Selection.
This procedure was adapted from several studies (Klemm et al., 2003; Ode et al., 2005; Hughes et al., 1998).

The range of metrics is determined to make sure it is broad enough to discern changes Percent > 10%; Richness > 5, Diversity >1, Evenness > 0.1
Area-based effects examined (Vannote et al., 1980). Examine relationship between metrics and catchment area None found – sites primarily 1st to 4th order
Examine responsiveness to primary urban gradient Visual inspection of wedge-shaped scatterplots, common in ecological relationships (Figures 10 and 11 in Thomson et al., 1996) 95th quantile regression used to identify strength of relationship Criteria for evaluation each plot described in Section 2.2.3
Metrics placed into correlation analysis to test for redundancy Metrics considered redundant if $r > 0.7$

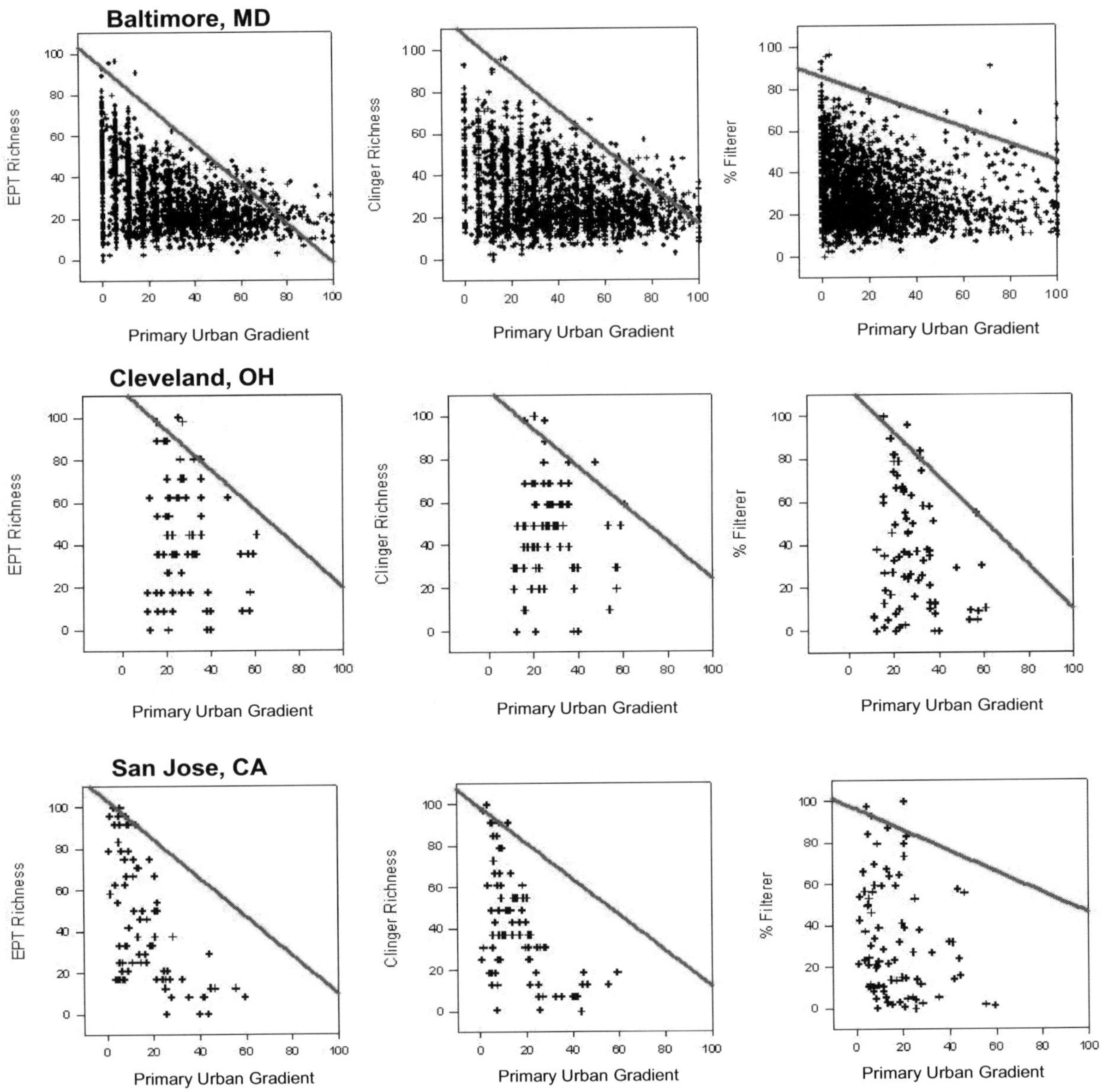

Figure 2-8. Examples of Metrics Plotted Against the Primary Urban Gradient for Each of the Three Study Regions. The line in each plot is the 95th quantile regression line.

The metric selection procedure resulted in a list of metrics selected within each region (Table 2-4). Metrics with a similar and strong response across regions were top candidates for use in the biological index. In addition to responding across regions, the final metrics were required to be: 1) easily measured, calculated and understood; 2) widely used by agencies; and 3) related to benthic macroinvertebrate assemblage structure and function. The three metric categories used here [composition/richness, trophic (functional feeding group), and habit] were meant to reflect characteristics of assemblage structure and function important in the TALU framework (U.S. EPA, 2005).

The final metrics selected for building the potential multimetric biological indices were: Ephemeroptera, Plecoptera, Trichoptera (EPT) percent and richness (structure), filterer percent and richness (function), and clinger percent and richness (habit). EPT was selected over the individual order richness metrics Ephemeroptera, Plecoptera, or Trichoptera, because EPT 1) had the strongest response to the primary urban gradient, 2) encompassed a larger range of taxa, and 3) was more broadly applicable (see also Carter & Fend, 2005). Filterer and clinger metrics were the only function and habit metrics that shared a strong response across regions.

2.2.2 Construct Candidate Multimetric Biological Indices

The six final metrics (percent EPT and EPT richness, percent filterer and filterer richness, and percent clinger and clinger richness) were standardized on a 0–100 scale using the 1st and 99th percentiles of the region-specific (i.e., East, West, Midwest) data distributions and assembled into indices by averaging their scaled values. The multimetric indices were constructed of every possible combination of richness and percent metrics with three in each index (Table 2-5). This procedure was consistent with earlier work in which combinations of percentage and richness metrics were used to construct multimetric indices (Karr, 1981).

Table 2-4. Best Performing Metrics in Each Study Region. Metrics in bold had a consistent and strong response across all three regions. Composition (%) and richness (taxa count) were used to characterize each metric category.

Study Region	Categories of Metrics		
	Structure	Function (FFG)	Habit
Baltimore, Maryland	**EPT** **Ephemeroptera** **Plecoptera** Trichoptera	**Filterer** Predator Scraper Shredder	**Clinger** Sprawler Swimmer
San Jose, California	**EPT** **Ephemeroptera** **Plecoptera** Trichoptera Taxa Richness	**Filterer** Scraper Shredder	**Clinger** Sprawler Swimmer
Cleveland, Ohio	**EPT** **Ephemeroptera** **Plecoptera**	**Filterer** Collector	**Clinger**

Table 2-5. Biological Indices Constructed Using Three Metrics Representing Categories of Structure, Function, and Habit.

Category	Index 1	Index 2	Index 3	Index 4
Structure	%EPT	EPT richness	%EPT	EPT richness
Function (FFG)	%Filterer	Filterer richness	Filterer richness	%Filterer
Habit	%Clinger	Clinger richness	%Clinger	Clinger richness

Category	Index 5	Index 6	Index 7	Index 8
Structure	%EPT	EPT richness	%EPT	EPT richness
Function (FFG)	%Filterer	Filterer richness	Filterer richness	%Filterer
Habit	Clinger richness	%Clinger	Clinger richness	%Clinger

2.2.3 Screen Potential Multimetric Indices to Identify Most Robust

The eight constructed biological indices (Table 2-5) were plotted against the primary urban gradient from Section 2.1. Data from all three regions were plotted together to examine their similarity. As mentioned previously, linear regression and Pearson correlations were not appropriate for characterizing these data because plots were primarily wedge-shaped. Quantile regression of the 95th quantile was used to characterize the upper boundary of the wedge-shaped plots. The quantitative criteria for evaluating each plot were:

- Slope of 95th quantile regression line must be approximately equal to -0.9 (when urban gradient = 100, Index must be greater than 10);
- The biological index must respond across the entire urban gradient (data points close to 100 on the urban gradient);
- All regions must have similar and overlapping responses; and
- The 95th quantile regression line must have a relatively high pseudo r-square value (this implies low variability around 95th quantile regression line).

The primary biological index that best met the above criteria was Index 6, composed of EPT richness, filterer richness, and percent clingers (Figure 2-9). This index was most responsive to the primary urban gradient and to the alternative gradients described in Section 2.1. The response envelope defined by the Baltimore data, which were far more numerous, also included San Jose and Cleveland. This suggests a very similar response for this index. Index 6 also responded consistently to the primary urban gradient across all three study regions, which suggests that it had broad geographic applicability and could potentially be used in areas other than those addressed here.

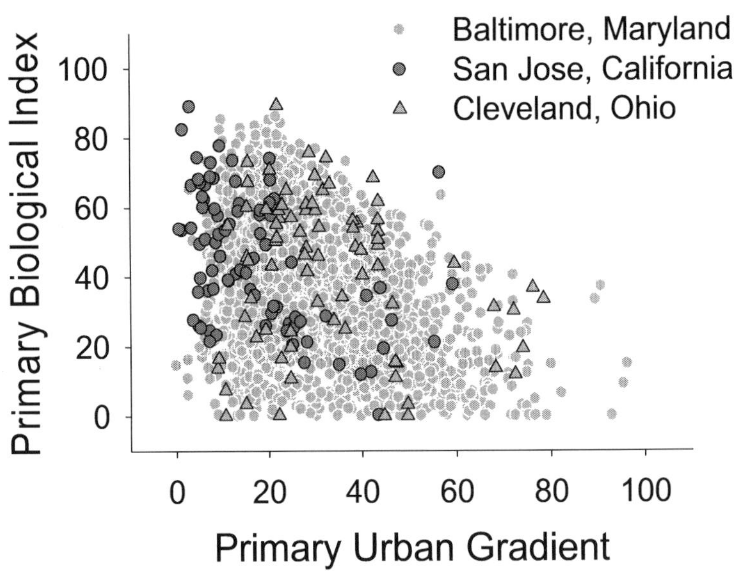

Figure 2-9. The Primary Biological Index (Index 6 – EPT Richness, Filterer Richness, and Percent Clingers) Plotted Against the Primary Urban Gradient for the Three Aggregated Study Regions.

2.2.4 Conclusions

The final primary biological index is used in Section 2.3 to define the biological potential along urban gradients. This index incorporates three metrics representing aspects of benthic macroinvertebrate assemblage structure and function and, thus, describes instream biological condition in a manner consistent with the TALU framework (U.S. EPA, 2005). Although this index may have national application, it should be noted that indicators with the best performance across regions are not necessarily the best indicators within each region. The implications of this research do not preclude the use of region-specific biological indicators. Many state water quality agencies have developed robust and calibrated indices for use within their jurisdictions.

2.3 Step 3: Define Biological Potential Along Urban Gradient(s)

The final step in the process is the determination of biological potential along urban gradients (Figure 2-10).

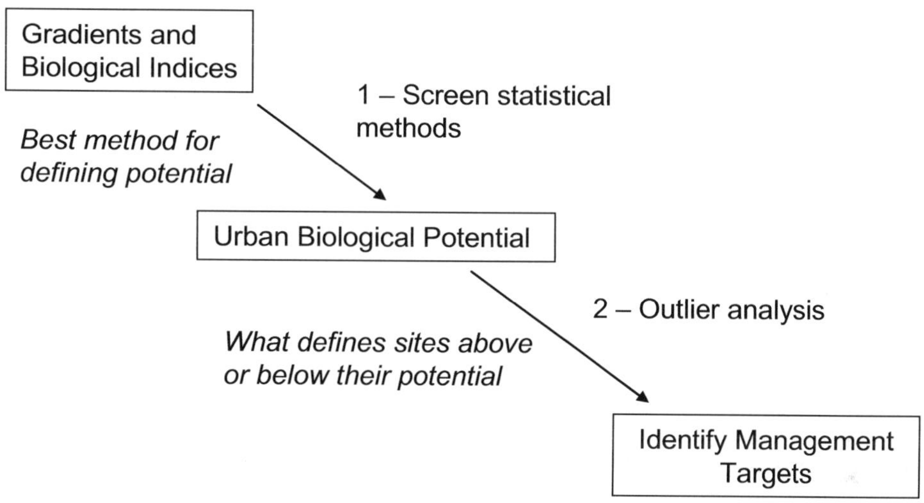

Figure 2-10. Process to Identify Biological Potential and Management Targets for Urban Streams.

As described earlier, the relationship between the urban gradients and biological response is defined as a wedge-shaped plot (e.g., Figure 1-1, and see Figure 2-9). These types of plots are common in biological assessment and monitoring, reflecting the multitude of natural and anthropogenic factors that interact to affect biological communities. The interesting aspect of these plots is the upper boundary—clearly the limit of maximum biological condition decreases as urbanization increases. It is this limit that defines the biological potential for a given level of urbanization. The term "potential" is used rather than "reference," which is reserved for a fixed biological condition in the absence of human disturbance—a reference condition for biological integrity (sensu Stoddard et al., 2006). The upper boundary—or potential—observed in the plots is most similar to a dynamic "least disturbed" condition, and reflects what a stream is capable of attaining based on current conditions.

The use of the biological potential approach is important for many reasons. First, it uses a "one yardstick" approach. Developing one scale for gauging all types of stream condition (from undisturbed to completely degraded) is preferable to using a variety of different scales, because it allows us to converse on equal terms regardless of the setting and does not lower the absolute bar for present and future management. Second, it is possible to define different capacities of stream condition under a variety of different landscapes that balance socio-economic realities and needs with environmental condition (e.g., TALU). These capacities are the potential described here. *Use of this approach recognizes that highly urbanized streams will not be comparable to streams in reference condition for biological integrity, but each urban stream should be able to attain a defined potential.* In addition, it is valuable to be able to prioritize restoration efforts on streams that can improve most relative to their potential. This approach provides a process for defining those streams with the most to gain from informed decision-making. The challenge lies in developing a tool to define that potential.

Aside from defining the biological potential of a stream at some point along a disturbance gradient, the next step is investigating why a stream is not meeting its potential. Even in urban catchments, there are large differences in the condition of streams as a result of differences in pollutant loads, riparian conditions, best management practices, etc. These all contribute to variability in condition for a given level of disturbance. An important management goal is, therefore, to define which aspects of landscape or instream condition that are part of the mosaic of urban land use are most associated with meeting biological potential. This information provides management targets and strategies for characterizing and improving the biological condition of those streams falling short of their potential.

2.3.1 Defining Potential

The process for defining biological potential consists of a statistical approach for describing the limit in biological condition with increasing urbanization. The upper boundary of the wedge-shaped plot defines that potential and there are few regression approaches that are appropriate. However, two main approaches were used in this study: quantile regression and regression through bin quantiles.

Quantile regression (Koenker & Bassett, 1978; Koenker & d'Orey, 1994; Koenker & Hallock, 2001) is a method for estimating lines through quantiles. Similar to how simple linear regression estimates a linear model around conditional mean estimates by minimizing the sums of squares of residuals, quantile regression uses a method for estimating models describing the relationship between conditional response quantiles and some predictor(s). The median quantile regression would represent a linear model through the central tendency, and this can be achieved by minimizing the sum of absolute errors. However, any other quantile function can also be estimated, generally by minimizing an asymmetrically weighted sum of errors (Koenker & Hallock, 2001). The result is a line defining the relationship between any independent predictor and the selected quantile of the dependent variable. The 75th and 90th are fit to the data and are selected to be consistent with general approaches for defining biological criteria. When using all the data in a distribution, as in this case, higher quantiles represent a better balance between errors associated with not making decisions of impairment when they exist (type II error) versus calling something impaired when it is not (type I error).

Regression through bins is a simpler approach. In this case, approximately equal sample-sized bins of observations for the dependent variable are arranged along the independent variable axis, a selected quantile of the bin is estimated, and a simple linear regression is estimated through the quantile estimates. For exploratory purposes, the 90th quantile is used for comparison to the quantile regression approach.

Quantile regressions yield visually more accurate potential lines through the upper boundary of the distribution of biological index values along the urban gradients than do the bin regressions (Figure 2-11). For this reason, only quantile regressions are used for the remainder of the analysis.

The biological potential is defined as the 90th quantile regression line. Very simply, one could designate the potential as the target and any site falling below it would be considered appropriate for management action. However, the 75th quantile regression line provides a lower boundary and the region between the two would be considered an allowable target, with sites below the 75th quantile considered as those sufficiently below potential to warrant management action. Use of this approach recognizes that there is some variability in potential that may reflect natural factors.

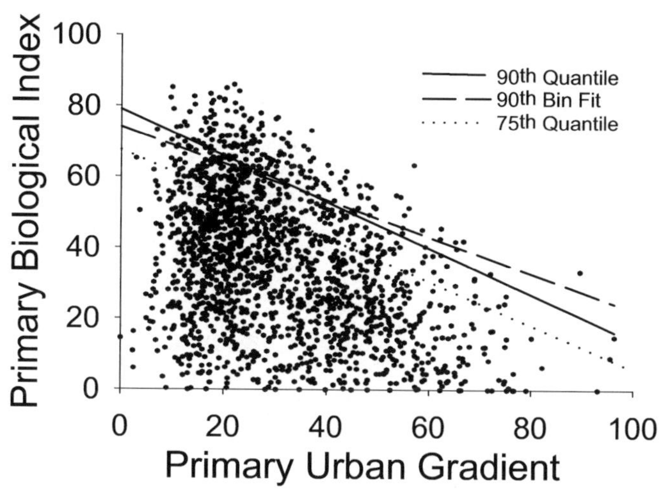

Figure 2-11. The 90th and 75th Quantile Regressions of the Primary Biological Index Against the Primary Urban Gradient. The regression through the bin 90th quantiles is also shown.

The number of sites falling above the 75th and 90th quantiles is shown in Table 2-6. This rather straightforward approach by definition means that 75 percent of the sites are below their potential, but this is solely on the basis of urbanization. If agriculture, mining, or other factors are taken into account, many of these streams may meet their potential for those settings. Many of those sites would be at or above the threshold for biological condition based on existing biocriteria thresholds, but this approach is not designed to supplant that. The goal of this approach is to be able to put those sites falling below the biocriteria thresholds into context relative to their potential given the landscape context, in this case, urbanization.

Table 2-6. The Number of Sites Above the 75th and 90th Quantiles Based on the Primary Urban Gradient.

Gradient	N>75th	N>90th	Total N
Primary Urban Gradient	469	187	1879

2.3.2 Outlier Analysis

It is evident that there are a number of sites above the 90th quantile (Figure 2-11). Having defined a biological potential, it is crucial to consider why some sites meet or exceed their potential, while others do not. The answer will enable managers to identify actual factors for use in developing actions to improve biological condition. For example, one factor associated with better than expected biological condition in urban streams is riparian condition (Yoder et al., 1999); another is the geomorphic habitat quality of the stream (Walters et al., 2003). The next analysis provides a process by which these and other factors might be identified to help guide management.

Two different approaches were compared for this analysis—one using all of the data together as a continuous analysis and one using the data split into urbanization bins. The first approach assumed that factors affecting streams were similar across the gradient from minimum to maximum urbanization. This may be true, but it is also possible that very urbanized catchments have landscape attributes substantially different from least urbanized catchments. For this reason, the second approach was used, which analyzed potential outliers within very urbanized catchments separate from those in least urbanized catchments. Data from the relationship between the primary urban gradient and biological index were used in this section to demonstrate the process.

In the first approach, all sites were scored relative to potential. This was defined as the difference between the observed score and that predicted from the equation for the 90th quantile regression. For the primary urban gradient—biological index relationship being used, the 90th quantile regression equation was:

$$\text{Index 6} = 79.0 - 0.65(\text{Primary Urban Gradient Score})$$

By subtracting the predicted scores from observed scores, sites above their potential were positive while those below it were negative. The range, in this example, was from -70 to +20. These values were used as the dependent variables in a multiple regression analysis using a few water quality and habitat parameters available for the Baltimore sites, as well as a variety of landscape parameters related to agriculture and natural land cover. Forward stepwise analysis (F-to-enter = 1.0) was used to select regression predictors. This analysis resulted in a model explaining 16 percent of the variance in the data (Table 2-7 and Figure 2-12). From the equation, sites that had better instream habitat (higher riffle frequency, better instream habitat scores, and lower embeddedness), lower conductivity, and lower catchment pasture cover score closer to their potential than others. This process could be used to prioritize the use of habitat restoration, water quality mitigation, and identify best management practices for pasturelands. Caution is advisable, as multiple linear regression in this context can not be used to identify causal relationships. But it suggests relationships that can be used to guide further inquiry in a site by site basis.

The binned scoring process separated sites along the primary urban gradient into bins, based on a similar sample size in each bin. For this analysis, again using the primary urban and biological index, there were six bins defined at primary urban gradient scores of 0–15, 15–25, 25–35, 35–45, 45–60, and >60. In this case, the 90th quantile was calculated in each bin and then sites scored as in the continuous method (i.e., by subtracting site scores from the 90th quantile score for each bin). Again, sites ranged in scores from negative (below potential) to positive (above potential). These bin scores were then placed into a multiple regression model as explained above for the continuous analysis. Because the process was the most important aspect of this discussion, only the first bin (0–15) analysis is discussed.

The equation for the first bin outlier scores explained 29% of the variance in the data (Table 2-7 and Figure 2-13). As with the continuous data, sites with better habitat quality (good channel flow and riffle frequency) and low conductivity scored closer to their potential. Surprisingly, even though the first bin had the lowest levels of urbanization, no agricultural land cover variables explained sites that scored below their potential. It may be that agricultural impacts were picked up by high conductivity and lower habitat quality or it may be that agriculture had less of an effect than presumed among the first bin sites. In either case, this result emphasizes the caution that should be taken in the interpretation of the results. This analysis approach is intended as a piece in the strength-of-evidence approach used to identify appropriate management strategies to mitigate biological impacts.

Table 2-7. Multiple Regression Equations and Statistics for the Continuous and Binned Outlier Analysis.

Continuous Scoring
Biological Index Score = -16 + 0.84 (Riffle Frequency) + 0.49 (Instream Habitat) - 6.3 (Embeddedness) - 6.1 (Log10Conductivity) -10 (Arcsine Pasture0.5)
R^2 = 0.16, Adjusted R^2 = 0.16, F=38.2, N=948
Binned Scoring - Bin 1 Model Example
Biological Index Score = 8.9 + 0.23 (Channel Flow Status) - 25 (Log_{10}Conductivity) + 0.51 (Riffle Frequency)
R^2 = 0.29, Adjusted R^2 = 0.27, F=18.28, N=135

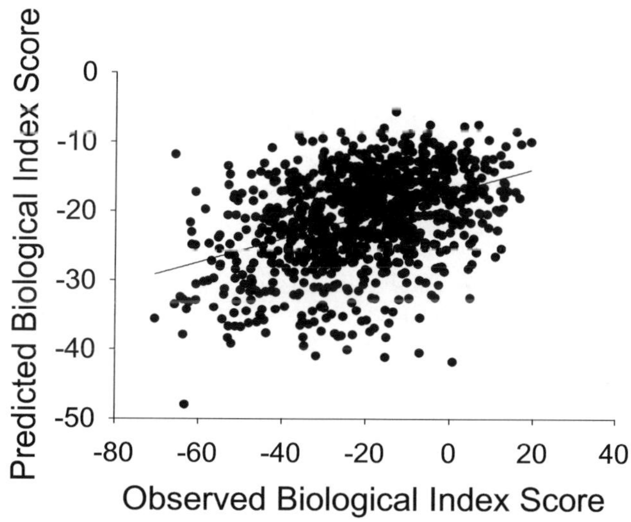

Figure 2-12. Observed vs. Predicted Values for the Continuous Outlier Analysis.

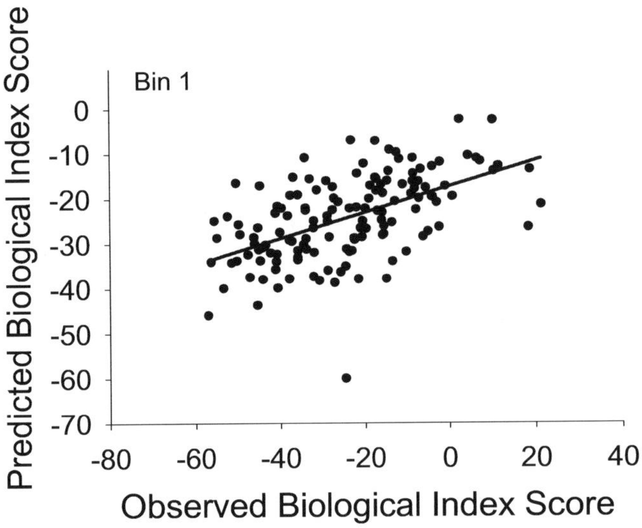

Figure 2-13. Observed vs. Predicted Values for the Binned Outlier Analysis for Bin 1 (0–15 primary urban gradient scores).

2.4 Synthesis and Summary of Scientific Findings

There are a multitude of landscape disturbances associated with urban areas and a wide variety of stressors originating from these sources. The effect of these stressors on nearby waterbodies is consistent and has recently been termed the "urban stream syndrome" (Meyer et al., 2005). Because of the complexity of the urban landscape and the variety of stressors that are produced, the restoration of urban systems is difficult for environmental managers and mitigation efforts often fail. Successful restoration of urban waters to a "least impacted" reference model may depend on the type, magnitude, and extent of stressors affecting the system. In nonurban catchments, least impacted reference sites have been used as reasonably attainable goals when linked to specific aquatic life uses. In urban catchments, as evidenced by the significant quantile regression relationships, restorable benchmarks may be more complex and vary along this threshold. Decisions on how to invest restoration dollars will rely on the existence of sound data on which to assess the feasibility of the proposed restoration.

The research presented here is related to three principal tasks: synthesizing the various impacts associated with urban development into one primary urban gradient, developing a sensitive a broadly applicable biological response indicator to urbanization, and defining an empirical estimate of the biological potential of streams at any point along the urban gradient. The biological potential here is developed as an attempt to move towards attainable biological conditions as embodied by the TALU framework for urban streams. This research is guided by a series of objectives, defined in Sections 1.1 and 1.2, and used here to synthesize the results.

2.4.1 Biological Indicators Provide a Useful Tool for Detecting Urban Stress

The strong relationships between the biological indicators we constructed and the disturbance gradients are evidence that these are useful tools for detecting urban stress. Biological assemblage indicators have been very successful in measuring biological deviation from minimally or least impacted conditions across the United States and Europe. In urban streams, existing biological indicators (e.g., Indices of Biological Integrity) frequently, if not characteristically, deviate from least disturbed conditions (see Chapter 1.0 and References). In this study, particular biological indicators respond specifically to quantified gradients of urban disturbance in a consistent manner suitable for using the responses for guiding management.

The goal of this project is not to develop new monitoring tools, but rather to use existing high-quality datasets provided by states, local entities (e.g., regional sewer authorities), and others to develop assemblage level indicators that are sensitive to urbanization. The intent is also to identify stressors and source indicators associated with urban landscape changes. It is important that the biological indicators detect a gradient of conditions along a disturbance gradient and fit a conceptual model of how biological assemblages may change with stress. This effort is consistent with U.S. EPA efforts to develop biological condition gradients (BCGs) along stressor-source gradients to assist states in refining their aquatic life uses (Davies & Jackson 2006; U.S. EPA, 2005). The objective of both efforts is to enhance the usefulness of aquatic life uses and water quality criteria, for water quality management. The focus of this work is on biological response to urban stressor gradients; however, widespread development of regional BCGs can provide a framework within which the biological and stressor gradients developed here should fit. The BCG is structured around the biological integrity goal of the Clean Water Act (CWA) and will be an essential tool for communicating the refinement of measurement endpoints or aquatic life use goals that may be implemented for urban waters.

2.4.2 The Biological Indicators Respond in a Useful Way to the Disturbance Gradients

To be useful in management of urban stressors the biological indicators examined here needed to respond in more than a pass/fail manner. It is clear from Sections 2.2 and 2.3 that the biological indicators reflected a reliable gradient of response to the primary urban gradient. The primary biological index shows strong threshold relationships with the primary urban gradient in each of the regions (e.g., Figure 2-9). These relationships show relatively sharp response surfaces that indicate the limiting effects of the disturbance gradient along a continuum of stress. At first glance, the relationship between ambient stressors and biological response indicators contains extensive scatter and variability. This observation is characteristic, however, of many complex relationships (e.g., economics) and many times a clear upper boundary exists. This relationship reflects the limitation of one or more stressors along the surface of the upper boundary with other factors more limiting below.

Economists have developed useful statistical tools to define this threshold, in particular, quantile regression (Cade & Noon, 2003). In this study, the upper boundary defines current limitations to aquatic biological potential. This potential was identified using quantile regression and defines an empirically defined target for biological condition along the urban gradient. Availability of a reliable biological indicator that can be linked directly to aquatic life uses or related goals is essential for effective management of urban stressors. Specifically, the urban biological indicator can be used in a process to set alternate aquatic life uses or interim measurement endpoints for restoration activities. The CWA, for example, requires a test of "feasibility of restoration" for setting less than CWA interim goals. This would be nearly impossible without an identified relationship between stressors and their sources and a biological response tied directly to an aquatic life use. Once the limitations to restoration are identified, then this information can form the basis for priority setting and the allocation of restoration resources.

2.4.3 General Applicability of the Urban Gradient-Biological Response Relationship

The urban biological response indicators behave similarly among eastern, midwestern, and western stream systems even when underlying measures varied substantially. However, indicators that worked better in some regions than others are probably related to variations in monitoring programs, sensitive organism groups, and even variations in urban disturbance gradients. This is not unexpected and local refinements to these urban indicators may be necessary for the most effective management of urban stressors in some regions and may provide additional advantages for entities that develop them. For example, biological indicators serve as reliable response indicators for stressor identification efforts and for tracking incremental progress of management activities. Section 2.4.2 of this report describes a process for understanding potential along the disturbance gradient and investigates why sites deviate from the potential represented by the threshold. Entities that develop the capacity to 1) calculate an urban disturbance gradient and 2) develop appropriate urban biological measures, will have established a basis to develop stressor identification procedures. For an institution developing such capacities internally, this may be a pre-requisite to overcoming inertia that inhibits biological data from being fully incorporated into local and regional decision-making. This would represent substantial progress beyond using biological data solely for assessment of impairment status.

2.4.4 Understanding Stressors Associated with Urban Intensity Gradients

Quantification of reliable urban gradients (Section 2.1) is a fundamental step in the process of understanding biological responses to urban stressors and untangling the urban stream syndrome. Recent efforts of other workers who quantified urban disturbance gradients that were correlated with biological conditions in many different settings are used as a springboard in this work. These approaches contain both stressor measures (e.g., chemical measures) and indicators associated with stressor source (e.g., land use and population). Source level data are, admittedly, one or more levels away from identifying the instream stressors that are likely the mechanisms of impairment. These source-level urban gradient measures are, however, strongly correlated with aquatic life impairment and provide a useful organizational framework within which instream stressors and management strategies can be identified.

In Section 2.1, the primary urban gradient based on population density, road density and urban LULC is developed. This gradient serves as the working model of urban intensity, being composed of parameters that represent urbanization as it is defined. Alternative urban gradients using stressor and stressor source variables are developed by selecting those that were most closely related to our primary urban gradient. These alternative gradients represent a variety of uses (i.e., source, stressor, mixed, and ease of cost or use). These alternative gradients are developed to present alternatives, strongly correlated with the primary gradient, that can be used in a variety of management situations (e.g., where the primary gradient variables are not available).

Once an association is found between urban gradients and biological responses, stressor identification procedures can be used to explore stressors at sites that deviate substantially from the potential (i.e., are not meeting their potential). The stressor-response relationships developed here are spatial descriptions because they rely on variation in stressor and response among sites and not across time. Where monitoring programs are ongoing and active best management practice implementation and adaptive management are practiced, temporal stressor patterns can be examined as data accumulate at sites over time (e.g., does the biology respond to abatement of specific stressors over time? Can the limiting factor change with best management practices? Figure 2-14). These data can be used to verify the potential generated from the spatial data and confirm or reject the limiting nature of various stressors in urban settings. Such feedback is an essential component of an active adaptive management program and can be used to identify the feasibility of reducing urban stressors and reducing the slope of threshold equations illustrated in Figure 2-14. These factors are critical components required to set alternate aquatic life uses.

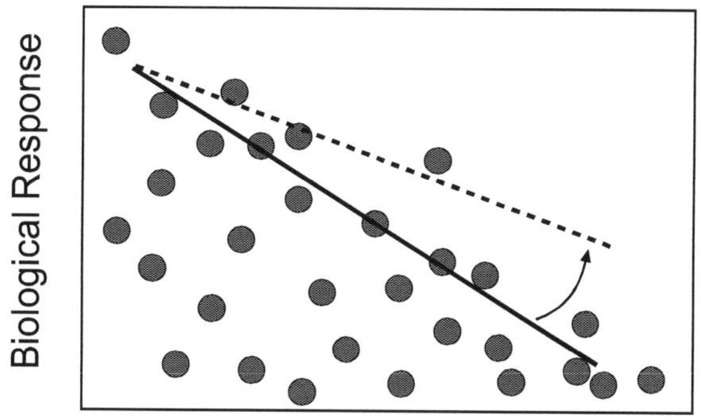

Figure 2-14. Illustration of How a Threshold Association May Change Over Time with Reduction of the Limiting Stressor.

2.4.5 Using Results to Help Direct and Prioritize Protection and Restoration Efforts

These results identify a biological potential that can be used to help prioritize both protection and restoration efforts. Identification of the stressors causing limitations along the threshold can be used to determine the feasibility or difficulty in restoring sites to their potential. If restoration effort is unfeasible under certain circumstances, alternate aquatic life uses may be more justified and a focus on preventing higher quality streams from accumulating these stressors gains higher importance.

Confidence in the relationships identified here is essential to effective and reasonable management of urban systems. The upper boundary relationships have been commonly observed where multiple stressors affect biological response. Use of quantile regression to describe these thresholds has been increasing and appears to be an effective tool for defining biological potential for these types of data (Cade et al., 1999; Cade & Noon, 2003; Pacheco et al., 2005). Both funding (e.g., stormwater, construction grants) and water quality management programs (e.g., total maximum daily loads [TMDLs], catchment protection and restoration) have priority-setting steps that are designed to funnel resources to protect and restore streams in an effective manner. The data used in this study are often absent from such efforts, but the approach described in this paper could be added to prioritization schemes to provide a direct measure of aquatic life use condition and biological potential.

CHAPTER 3.0

IMPLEMENTATION – LINKING SCIENTIFIC PROCESS AND AQUATIC LIFE USE

Aquatic life uses and their associated water quality criteria are the primary pathway for setting protection and restoration goals for waterbodies—in this case rivers and streams. The key is using appropriate tools for applying water quality standards in complex urban environments, which in turn will have great benefits for cost effective and environmentally sound administration of permits and other CWA programs that affect WERF subscribers. The biological indicators, urban benchmarks, and analytical tools demonstrated in this study have practical applications in a water quality standards framework, particularly through refinement of aquatic life uses and associated water quality criteria.

3.1 The Use of Stressor Gradients to Set Alternate Management Goals for Different Streams in an Urban Setting

Most states do not yet have TALU (Davies & Jackson, 2006); however the types of relationships demonstrated in this study provide the basis for consideration of alternative management of streams along urban gradients. Confidence in the urban benchmark gradients and the biological responses explored here is necessary to justify alternate management strategies to regulators. This study demonstrates that uncertainty is reduced by the capacity to quantify these relationships though the use of quantile regression. For example, in this study Index 6 had the strongest relationship with the primary urban gradient (Section 2.2.3). Further, these relationships fit within conceptual models of biological change and should be compatible with national efforts to construct BCGs that are the foundation for efforts to refine aquatic life uses. (U.S. EPA, 2005). The next step would be to demonstrate and justify how these indicators relate to the policy foundations of the various water quality programs.

The phenomenon of the urban stream syndrome is real and limitations to biological condition are highly predictable along an urban gradient. The stressor-response relationships observed in all three regions are technically convincing (Sections 2.3 and 2.4). Most urban water quality regulation (e.g., via stormwater regulations) relies on prescriptive approaches that are not necessarily tightly linked to measures of ecological accountability, but more often to administrative goals. Such approaches are easy to implement and are administratively sound, but may be expensive if the ultimate aquatic life goals are not met or are unattainable. The approach described here allows dischargers in heavily urban areas, for example, to assess ecological condition below facilities in the existing context (i.e., the biological potential given the urban setting) rather than a reference condition based on forested conditions. Upon determination of the potential of a particular waterbody, specific management activities can be installed to help the waterbody attain its biological potential.

The few states that have incorporated refined aquatic life uses already had established credible long-term monitoring programs in 1980s and 1990s before adoption of refined aquatic life uses (e.g., Ohio, Maine) or assessment endpoints that reflected gradients based on existing conditions (e.g., Texas; see U.S. EPA, 2005). For these programs long-term monitoring efforts were relatively stable, which allowed "lessons-learned over time" to filter into management actions. In Ohio, numerical biocriteria and TALU were established in 1990; however more complete integration into management programs through revised 208 Water Quality Plans has been more recent (Ohio Environmental Protection Agency, 2006). Adoption of active adaptive management (sensu Allan & Curtis, 2005), which embraces monitoring data and analysis as represented by this study could accelerate this process of data integration to facilitate informed decision-making. Providing an approach to quantify the urban gradient and the resulting biological response should enhance the acceptance of alternate management goals.

3.2 Application of Results to Specific Urban Stream Management Efforts

Aquatic life uses and associated water quality criteria have a pervasive influence on most water quality programs because they embody the goals of programs and the criteria to attain these goals. Refining aquatic life uses, which adjusts the goals and potentially the criteria necessary to meet these goals can have substantial influence on program areas and thus provide a rational strategy for managing urban streams. Major program areas that would benefit from the refinement of aquatic life goals are summarized in Table 3-1 although benefits of refined goals would extend to most applications of water quality management.

Municipalities should be able to apply this process to determine biological potential of their streams and urban watersheds. This type of assessment would provide evidence of possible mitigation efforts and priorities of management decisions.

3.3 Recommendations for Biological Monitoring Programs and Urban Catchments Studies

The biological indicators developed here are not replacements for statewide assessment tools such as multimetric indices or multivariate approaches that are in current use. These tools are typically designed to assess whether sites are comparable to least impacted reference sites and respond to a broad range of potential stressor types. Rather than use a composite index, a subset of metrics is selected that is especially sensitive to urban disturbance to provide the most precise response to the urban benchmark gradients. It is important, however, that the urban gradient described here can be linked, at least conceptually, to a more comprehensive BCG at a statewide or regional geographic scale (see Chapter 1.0). For WERF subscribers, using existing indicators already calibrated and incorporated in a monitoring program can increase acceptability of results. If existing indicators are not sufficiently robust, however, urban studies may need to employ a more sophisticated approach.

The existence of a robust monitoring program at a state level should increase the acceptance of urban gradients demonstrated in this study. Experiences from Maine and Ohio on implementing TALU, a major potential use of this work, identified a number of detailed milestones that were met before deriving tiered uses (U.S. EPA, 2005); they include:

- establishing a conceptual foundation for the biological measures,
- merging scientific and policy foundations (i.e., link management objectives with water quality standards),
- establishing the monitoring program, and
- developing/validating quantitative (biological) thresholds.

The experiences of states such as Maine and Ohio suggest that the biological indicators, urban disturbance gradients, and analytical procedures demonstrated in this research are basic to any programmatic effort of water resource management.

Table 3-1. Selected Program Areas Where the Urban Biological Indicators and Urban Gradient Construction in This Study Can be Used to Influence Management Decisions Through Aquatic Life Use Refinement.

Program Area	Role of Biosurvey & Stressor/Source Data	Description
1. Water quality standards	1a. Refined designated uses	A robust biosurvey is a prerequisite for TALUs. For uses below the interim aquatic life use, the approach here provides the information required to quantify aquatic potential along measured disturbance gradients and a process to define and assess feasibility of restoration required under the CWA.
	1c. Refined and validated water quality criteria	Biological and stressor data can be used in combination to develop and validate water quality criteria, especially for nontoxic parameters that are not amenable to laboratory studies (e.g., habitat) or for which aquatic life criteria do not exist (e.g., nutrients, specific conductance, etc.). Where tiers are developed criteria may vary with the sensitivity of the assemblages.
2. TMDLs	2. Enhanced listing and de-listing criteria; derivation of TMDL stressor targets	The National Academy of Sciences (National Research Council, 2001) maintains that indicators closest in position to a use are the most reliable indicators of attainment or impairment of designated use. For aquatic life uses, this is largely biological assemblage data. These data can also be used to justify de-listing once restoration takes place. Furthermore, in a TALU framework, the biological indicators used to develop tiers, would become the assessment endpoints. In lieu of existing water quality criteria linked to uses, linkages between biological endpoints that define a use and stressors can be used to derive stressor targets (e.g., Pacheco et al 2005) appropriate to the position on the disturbance tiers
3. NPDES	3a. CSOs and other point sources	Biological and stressor data can be used to quantify the impact of individual CSOs to biological condition in an urban context and provide a basis for priority setting in relation to control plans and, through TALU, reasonable goals.
	3b. Stormwater (non-CSO)	Many stormwater restoration projects are cookbook, at most focusing on chemical impacts or water quantity. Furthermore, they are not well integrated with aquatic life uses and impairments even though stormwater-related impairments are a major cause of TMDL listings for states. Besides advantages derived from alternative aquatic life uses, an approach based on sound stressor-response data could improve the efficacy of stormwater efforts by a focus on identifying limiting stressors and driving feasible monitoring and restoration efforts. In addition, the impact of stormwater on biological condition has to be gauged within the context of a realistic potential, given the urban setting.

3.4 Critical Elements for an Urban Monitoring Program

Development of an adequate urban monitoring and assessment program relies on knowledge of the urban gradient, calibrated bioassessments tied to aquatic life uses, knowledge of aquatic life potential, and the ability to integrate these elements into water quality management program (U.S. EPA, 2003). This research digest outlines core environmental indicators including multiple organism groups for biological indicators, key chemical indicators (e.g., dissolved oxygen, nutrients, conductivity, pH, etc.), habitat, flow measures, and land use measures (U.S. EPA, 2003).

There are a number of important elements of biological monitoring data and programs that will affect their suitability for use in urban catchments (for more details regarding monitoring program elements, see Barbour & Yoder, 2006). The rigor of the methods and resultant data is important because increased rigor translates to decreased uncertainty in an assessment and/or management action. Aquatic life use refinement will require public review and will involve multiple stakeholders; scientific rigor may reduce resistance to refined uses. Barbour and Yoder (2006) examined the technical elements of monitoring programs and categorized monitoring programs into four levels of increasing capability and rigor based on a suite of 12 characteristics. Level 1 programs are essentially educational and not sufficient for characterizing urban disturbance gradients. Level 2 programs are pass/fail approaches and are also inadequate to distinguish gradients of biological response along one or more urban disturbance gradients. Table 3-2 illustrates some important technical elements (Barbour & Yoder, 2006) of an urban monitoring program (i.e., levels 3 and 4) that would be needed to fully implement the approaches outlined in this report.

3.5 Benefits and Challenges to Integrating Monitoring into "Everyday" Management Efforts in Urban Ecosystems

Despite the potential promise of integrating biological monitoring and stressor identification efforts into management of urban stream systems, there are still significant challenges and disincentives to such a program. Acceptance in states and counties with active monitoring programs should be easier than in those with rudimentary programs, because decision-makers will have more familiarity with the use of monitoring data in other regulatory applications (e.g., TMDLs).

There will be some initial challenges in communicating the results of complex statistics such as quantile regression, however such techniques help make sense of very complex urban ecosystems in ways that were perceived as nearly intractable. Biological response indicators are now widely accepted as the best way to measure achievement of aquatic life uses, so refined biological indicators fit into most states monitoring frameworks. Uncertainty generally causes regulatory agencies to be conservative in management of the environment for good reason, because their charge, in relation to aquatic life uses, is to protect and resource these uses. Adequate monitoring data and adequate data to construct benchmark gradients are fundamental to reducing uncertainly related to development of alternate aquatic life uses or alternate ways to manage existing aquatic life uses.

Table 3-2. Selected Characteristics of Level 1–2 vs. Level 3–4 Monitoring Programs and Application to the Characterization of Urban Systems.

Monitoring Program Attribute	Level 1-2		Level 3-4	
	Characteristics	**Application to Urban Assessment**	**Characteristics**	**Application to Urban Assessment**
Taxonomic Resolution	Family level	Pass-fail, Unlikely to classify more than 2–3 categories of condition	Genus/species level	Multiple classes can be distinguished, can support tiered uses or multiple restoration endpoints; Development of numeric criteria in Level 4 may result in the ability to distinguish a clearer gradient
Reference Sites	Fewer, less quantified criteria	Insufficient to classify aquatic potential along disturbance gradients	More reference sites, ability to distinguish among reference levels	For this procedure substantial sites needed to characterize threshold
Multiple Indicators	One	Lack of taxonomic resolution trumps use of multiple groups	One-two	Two organism groups would be ideal to identify organisms most limited by urban disturbances under differing scenarios and multiple scales; was not investigated in this study
Spatial Sampling	May be limited	May under-represent certain stressors or conditions	Usually intensive	Level 3 may under-represent certain categories of urban stress, Level 4 should cover most stress categories and provide a strong gradient
Diagnosing Problems	Limited to distinguishing relative severe or relatively unimpaired conditions	Typically only useful in identify very good and very severe urban impacts	Level 3 can provide categorical response indicators; Level 4 can provide complementary lines of evidence	Provides necessary resolution to diagnose problems. Level 4 could provide additional insight via additional lines of evidence (e.g., second organism group, more representative data coverage)

Once an urban potential is accepted, however, there are still important challenges to establishment of refined aquatic life uses. Key questions that must be addressed before establishment of alternate aquatic life uses include identification of limiting stressors, and feasibility of stressor remediation. There will also be issues related to the choice of threshold as an aquatic life goal (e.g., 75th vs. 90th quantile, etc.). The outlier analysis in Section 2.3.2 provides the approach for identifying the specific stressors limiting aquatic life and can provide information to help choose a management strategy. Once stressors are identified, then other information can be incorporated and the feasibility, cost, and likelihood of restoration can be addressed. The outlier analyses within bins of the primary urban gradient can be a powerful tool for developing reasonable and protective restoration options. The gradient itself is not likely to change because population density, road density, and urban land use/land cover will be relatively fixed under most circumstances. Within a range of an urban benchmark stressor values vary and are responsible for much of the biological variation within that benchmark range (e.g., 15–30%). For example, riparian habitats could be enhanced or restored, and may be sufficient for reaching the biological potential. Also, the "connectedness" of impervious surfaces to receiving streams is critical to the degree of biological impacts (Walsh et al., 2006), as is the pattern of urban development. These and other predictors defining the geometry of urbanization could be explored for their effect on biological condition relative to potential using the outlier analysis process and would provide valuable areas for management intervention to improve biological condition. Using this tool in this way, iterative examination of data from future monitoring can be used to set and refine estimates of aquatic life potential and to identify stressors amenable to restoration.

Successful implementation of the tools constructed in this study is beyond simply acquiring the scientific capacity to perform the analyses and collect the data. Indeed, many organizations may already have suitable scientific expertise. Institutions need to identify the specific program areas (see Table 2-1) where these data and approaches can be used and explicitly incorporate ecological endpoints, such as the urban indicators developed here, as measurable program goals. The threshold relationships identified in this study are compelling and promise more refined management of urban stream systems that will be both protective of aquatic life potential and yet be useful to efficiently prioritize scarce resources.

References

Allan, C.; Curtis, A. Nipped in the Bud: Why Regional Scale Adaptive Management Is Not Blooming. *Environ. Mgmt.* **2005**, *36*(3), 414–425.

Baker, B.B.; Richards, R.P.; Loftus, T.T; Kramer, J.W. A New Flashiness Index: Characteristics and Applications to Midwestern Rivers and Streams. *J. Am. Water Resour. Assn.* **2004**, *40*(2), 503–522.

Barbour, M.T. The Re-invention of Biological Assessment in the U.S. *Human Ecol. Risk Assess.* **1997**, *3*, 933–940.

Barbour, M.T.; Diamond, J.; Fowler, B.; Gerardi, C.; Gerritsen, J.; Snyder, B.; Webster, G. *An Assessment of the Status, Usefulness, and Resource Requirements for Implementation of Biocriteria Water Quality Standards*; Project No. 97-IRM-1; Water Environment Research Foundation: Alexandria, Va., 1999a.

Barbour, M.T.; Gerritsen, J.; Snyder, B.D.; Stribling, J.B. *Rapid Bioassessment Protocols for Use in Streams and Wadeable Rivers: Periphyton, Benthic Macroinvertebrates, and Fish*; 2nd Edition; EPA/841-B-99-002; U.S. Environmental Protection Agency, Office of Water: Washington, D.C., 1999b.

Barbour, M.T.; Yoder, C. *Critical Elements of a Bioassessment Program* (Draft). Prepared by Tetra Tech and Midwest Biodiversity Institute for U.S. Environmental Protection Agency: Washington, D.C., 2006.

Brown, M.T.; Vivas, M.B. Landscape Development Intensity Index. *Environ. Monit. Assess.* **2005**, *101*, 289–309.

Cade, B.S.; Terrell, J.W.; Schroeder, R.L. Estimating Effects of Limiting Factors with Regression Quantiles. *Ecology* **1999**, *80,* 311–323.

Cade, B.S.; Noon, B.R. A Gentle Introduction to Quantile Regression for Ecologists. *Frontiers Ecol. Environ.* **2003,** *1*, 412–420.

Carter, J.L.; Fend, S.V. Setting Limits: The Development and Use of Factor-Ceiling Distributions for an Urban Assessment Using Macroinvertebrates. In *Effects of Urbanization on Stream Ecosystems*; Brown, L.R.; Gray, R.H.; Hughes, R.M.; Meador, M.R., Eds.; Symposium 47; American Fisheries Association: Bethesda, Md., 2005.

Davis, W.S. Biological Assessment and Criteria: Building on the Past. In *Biological Assessment and Criteria. Tools for Water Resource Planning and Decision Making*; Davis, W.S.; Simon, T.P., Eds.; Lewis Publishers: Boca Raton, Fla., 1995; pp 15–29.

Davies, S.P.; Jackson, S.K. The Biological Condition Gradients: A Descriptive Model for Interpreting Change in Aquatic Ecosystems. *Ecol. Appl.* **2006**, *16*(4):1251–1266.

Gerritsen, J.; Carlson, R.; Charles, D.L.; Dycus, D.; Faulkner, C.; Gibson, G.R.; Kennedy, R.H.; Markowitz, S.A. *Lake and Reservoir Bioassessment and Biocriteria. Technical Guidance Document;* EPA 841-B-98-007; U.S. Environmental Protection Agency, Office of Water, Washington, DC., 1998.

Hughes, R.M.; Kaufmann, P.R.; Herlihy, A.T.; Kincaid, T.M.; Reynolds, L.; Larsen, D.P. A Process for Developing and Evaluating Indices of Fish Assemblage Integrity. *Can. J. Fisheries Aq. Sci.* **1998**, *55,* 1618–1631.

Karr, J.R.; Chu, E.W. *Restoring Life in Running Waters: Better Biological Monitoring*. Island Press: Washington, D.C., 1999.

Karr, J.R. Assessment of Biotic Integrity Using Fish Communities. *Fisheries* **1981**, *6*, 21–27.

Klemm, D.J.; Blocksom, K.A.; Fulk, F.A.; Herlihy, A.T.; Hughes, R.M.; Kaufmann, P.R.; Peck, D.V.; Stoddard, J.L.; Thoeny, W.T.; Griffith, M.B.; Davis, W.S. Development and Evaluation of a Macroinvertebrate Biotic Integrity Index (MBII) for Regionally Assessing Mid-Atlantic Highland Streams. *Environ. Mgmt.* **2003**, *31*, 656–669.

Koenker, R.; Bassett, G. Regression Quantiles. *Econometrica* **1978**, *46*, 33–50.

Koenker, R.W.; d'Orey, V. Remark on Alg. AS 229: Computing Dual Regression Quantiles and Regression Rank Scores. *Appl. Stat.* **1994**, *43,* 410–414.

Koenker, R.; Hallock, K.F. Quantile Regression. *J. Econ. Perspect.* **2001**, *15*, 143–156.

Lindsey, A.; Swietlik, W.F.; Hall, W.E. Effects of Watershed Development and Management on Aquatic Ecosystems - EPA's Perspective. In *Effects of Watershed Development and Management on Aquatic Ecosystems*; Roesner, L.A., Ed.; American Society of Civil Engineers: New York, **1997**; pp 4–16.

McMahon, G.; Cuffney, T.F. Quantifying Urban Intensity in Drainage Basins for Assessing Stream Ecological Conditions. *J. Am. Water Resour. Assoc.* **2000**, *36*(6), 1247–1261.

Meyer, J.L.; Paul, M.J.; Taulbee, W.K. Stream Ecosystem Function in Urbanizing Landscapes. *J. N. Am. Benthol. Soc.* **2005**, *24*(3), 602–612.

National Research Council. *Assessing the TMDL Approach to Water Quality Management.* National Academies Press: Washington, D.C., 2001.

Ode, P.R.; Rehn, A.C.; May, J.T. A Quantitative Tool for Assessing the Integrity of Southern Coastal California Streams. *Environ. Mgmt.* **2005**, *35*(4), 493–504.

Ohio Environmental Protection Agency. *208 Plan Prescriptions for Water Quality Protection within the Big Darby Creek Watershed Applicable to Portions of: Champaign County, Franklin County, Logan County, Madison County, Pickaway County, Union* County; Draft State WQM Plan. Ohio Environmental Protection Agency: Columbus, Ohio. 2006; Appendix 9-3.

Pacheco, M.A.; McIntyre, D.O.; Linton, T.K. Integrating Chemical and Biological Criteria. *Environ. Toxicol. Chem.* **2005,** *24,* 2983–2991.

Paul, M.J.; Meyer, J.L. Streams in the Urban Landscape. *Annu. Rev. Ecol. System.* **2001**, *32*, 333–365.

Paulson, C.L.; Reeves, S.; Canton, S.; Moore, T. *Distinguishing the Relative Influence of Habitat and Water Quality on Aquatic Biota;* Report No. 98-WSM-1; Water Environment Research Foundation: Alexandria, Va., 2001.

Richter, B.D.; Baumgartner, J.V.; Powell, J.; Braun, D.P. A Method for Assessing Hydrologic Alteration within Ecosystems. *Conserv. Biol.* **1996,** *10*(4),1163–1174.

Rosenberg, D.M.; Resh, V.H., Eds. *Freshwater Biomonitoring and Benthic Macroinvertebrates.* Chapman & Hall, Inc.: New York, 1993.

Schueler, T.R. 1994. The Importance of Imperviousness. *Watershed Protect. Techniq.* **1994**, *1*, 100–111.

Stoddard, J.L.; Larsen, D.P.O.; Hawkins, C.P.; Johnson, R.K.; Norris, R.H. Setting Expectations for the Ecological Condition of Streams: The Concept of Reference Condition. *Ecol. Applic.* **2006,** *16*(4), 1267-1276.

Tate, C.M.; Cuffney, T.F.; McMahon, G.; Giddings, E.M.P.; Coles, J.F.; Zappia, H. Use of an Urban Intensity Index to Assess Urban Effects on Streams in Three Contrasting Environmental Settings. In *Effects of Urbanization on Stream Ecosystems*; Brown, L.R.; Gray, R.H.; Hughes, R.M.; Meador, M.R., Eds.; Symposium 47; American Fisheries Association: Bethesda, Md., 2005; pp 291–315.

Tetra Tech, Inc. *Flow Time Series Estimation Tool.* Water Environment Research Foundation: Alexandria, Va., 2006.

Thomson, J.D.; Weiblen, G.; Thomson, B.A.; Alfaro, S.; Legendre, P. Untangling Multiple Factors in Spatial Distributions: Lilies, Gophers, and Rocks. *Ecology* **1996**, *77*(6), 1698–1715.

U.S. Environmental Protection Agency (U.S. EPA). *Use of Biological Information to Better Define Designated Aquatic Life Uses in State and Tribal Water Quality Standards: Tiered Aquatic Life Uses. Tiered Aquatic Life Uses Document*; Draft; EPA-822-R-05-001; Health and Ecological Criteria Division (4304T), Office of Science and Technology, Office of Water, U.S. EPA: Washington, D.C., 2005.

U.S. EPA. *Elements of a State Water Monitoring and Assessment Program,* EPA 841-B-03-003; Assessment and Watershed Protection Division, Office of Wetlands, Oceans and Watershed, U.S. EPA: Washington, D.C. 2003.

U.S. EPA. *Estuarine and Coastal Marine Waters: Bioassessment and Biocriteria Technical Guidance;* EPA 822-B-00-024; U.S. Environmental Protection Agency, Office of Water, Washington, D.C., 2000.

U.S. EPA. *Biological Criteria: Technical Guidance for Streams and Rivers*; EPA 822-B-94-001; Office of Science and Technology, U.S. EPA: Washington, D.C., 1996a.

U.S. EPA. *Watershed Approach Framework*; EPA 840-5-96-001; Office of Water, U.S. EPA: Washington, D.C., 1996b.

U.S. EPA. Biological Criteria: National Program Guidance for Surface Waters; EPA 440-5-90-004; Office of Water Regulations and Standards, U.S. EPA: Washington, D.C., 1990.

Vannote, R.L.; Minshall, G.W.; Cummins, K.W.; Sedell, J.R.; Cushing, C.E. The River Continuum Concept. *Can. J. Fish. Aquat. Sci.* **1980**, *37*(1), 130–137.

Walsh, C.J.; Roy, A.H.; Feminella, J.W.; Cottingham, P.D.; Groffman, P.M.; Morgan, R.P., II. The Urban Stream Syndrome: Current Knowledge and the Search for a Cure. *J. N. Am. Benthol. Soc.* **2005**, *24*(3), 706–723.

Walters, D.M.; Leigh, D.S.; Bearden, A.B. Urbanization, Sedimentation, and the Homogenization of Fish Assemblages in the Etowah River Basin, USA. *Hydrobiologia* **2003**, *494*, 5–10.

Yoder, C.O.; Miltner, R.J.; White, D. Assessing the Status of Aquatic Life Designated Uses in Urban and Suburban Watersheds, In *Proceedings, National Conference on Retrofit Opportunities for Water Resource Protection in Urban Environments;* EPA-625-R-99-002; Everson, A.; Minamyer, S.; Dye, J.; Heimbrock, P.; Wilson, S., Eds.; U.S. EPA: Washington, D.C., 1999; pp 16–28.

Yoder, C.O.; Rankin, E.T. Biological Response Signatures and the Area of Degradation Value: New Tools for Interpreting Multimetric Data. In *Biological Assessment and Criteria. Tools for Water Resource Planning and Decision Making*; Davis, W.S.; Simon, T.P., Eds.; Lewis Publishers: Boca Raton, Fla., 1995; pp 263–286.

APPENDIX A

DATA FROM URBAN GRADIENT DEVELOPMENT

Table A-1. Correlation Coefficients (*r* Values) of Chemistry, Habitat, Hydrologic, and Landscape Parameters Versus Primary Urban Gradient (Scaled and Averaged Population Density, Road Density and % Urban LULC).
Variables that were highly correlated with the primary urban gradient were considered for incorporation into alternative urban gradients (Section 2.1).

Chemistry	r-value
Acid Neutralizing Capacity	**0.36**
Ammonia	0.10
Chlorides	**0.65**
Conductivity	**0.59**
Dissolved Organic Carbon	**0.16**
Nitrate	-0.06
Orthophosphate	**-0.20**
pH (LAB)	**0.18**
Sulfate	**0.39**
Acid Neutralizing Capacity Resids	**0.35**
Ammonia Resids	0.06
Chlorides Resids	**0.55**
Conductivity Resids	**0.46**
Dissolved Organic Carbon Resids	-0.06
Nitrate Resids	**0.37**
Orthophosphate Resids	0.03
pH (LAB) Resids	**0.31**
Sulfate Resids	**0.27**

Habitat	r-value
Bank stability	0.01
Channel alteration	**-0.11**
Channel flow status	-0.05
Epifaunal substrate/available cover	**-0.15**
Frequency of riffles (or bends)	**-0.11**
Instream habitat	**-0.15**
%Embeddedness	**0.10**
Pool variability	**-0.11**
Riparian veg zone width	**-0.24**
Sediment deposition	**-0.12**
%Shading	0.01
Vegetative protection	**-0.19**
Velocity/depth regime	**-0.12**

Hydrology	r-value
Oct. Mean	**0.29**
Nov. Mean	**0.27**
Dec. Mean	**0.12**
Jan. Mean	**0.05**
Feb. Mean	-0.01
Mar. Mean	**-0.04**
Apr. Mean	**-0.12**
May. Mean	**0.07**
Jun. Mean	**0.14**
Jul. Mean	**0.40**
Aug. Mean	**0.52**
Sep. Mean	**0.52**
1-Day Min	**-0.08**
3-Day Min	**-0.07**
7-Day Min	**-0.06**
30-Day Min	**0.08**
90-Day Min	**0.31**
1-Day Max	**0.39**
3-Day Max	**0.33**
7-Day Max	**0.24**
30-Day Max	**0.12**
90-Day Max	**0.08**
Timing Min	-0.02
Timing Max	**0.48**
Low Pulse Count	**0.42**
Low Pulse Duration	**-0.30**
High Pulse Count	**0.40**
High Pulse Duration	**-0.28**
Rise Rate	**0.49**
Fall Rate	**-0.61**
R-B Flashiness	**0.56**

Landscape	r-value
%Bachelors Degress	**-0.05**
Children/Woman	-0.04
Proportion of Population >$100K	0.04
Per Capita Income	**-0.07**
House Income	-0.03
Housing Unit Density	**0.91**
Proportion of Housing Units Occupied	0.02
% Renter Occupied	0.04
Age of Housing Unit	0.03
Toxic Release Inventory (TRI) Density	**0.15***
Industrial Facilty Discharge (IFD) Density	**0.21***
Permit Compliance System (PCS) Density	**0.24***
Dam Density	**0.18***
Low Density Residential	**0.06**
High Density Residential	**0.79**
Commercial/Industrial	**0.61**

*due to outliers

Table A-2. Multiple Regressions of Primary Urban Gradient (Scaled and Averaged Population Density, Road Density, and %Urban LULC) Against Chemistry, Hydrologic, and Landscape Parameters. Variables that show high significance showed low (*P* value) and strong correlation (*r*-square) were considered for incorporation into the alternative urban gradients (Section 2.1).

	Beta	Std.Err. of Beta	B	Std.Err. of B	t(352)	p-level
Intercept			-52.42	9.35	-5.61	0.00
L_Conductivity	0.76	0.06	43.24	3.47	12.46	0.00
L_Nitrate	-0.22	0.04	-6.85	1.36	-5.04	0.00
L_Orthophosphate	-0.16	0.04	-4.45	1.16	-3.83	0.00
L_Acid Neutralizing Capacity	-0.13	0.06	-8.80	4.28	-2.06	0.04

	Step +in/-out	Multiple R	Multiple R-square	R-square change	F - to entr/rem	p-level	Variables included
L_Conductivity	1	0.59	0.35	0.35	194.55	0.00	1
L_Nitrate	2	0.64	0.41	0.06	35.74	0.00	2
L_Orthophosphate	3	0.66	0.44	0.03	17.28	0.00	3
L_Acid Neutralizing Capacity	4	0.67	0.45	0.01	4.23	0.04	4

	Beta	Std.Err. of Beta	B	Std.Err. of B	t(2297)	p-level
Intercept			50.34	3.65	13.81	0.00
L_BAKERSF	1.21	0.06	109.03	5.38	20.26	0.00
G4_HPD	0.66	0.03	9.54	0.46	20.75	0.00
G2_7DMIN	0.26	0.02	25642.14	1960.35	13.08	0.00
G4_HPC	0.08	0.02	0.34	0.11	3.18	0.00
G2_1DMAX	-0.10	0.03	-106.60	33.40	-3.19	0.00
G4_LPC	-0.07	0.02	-0.26	0.08	-3.15	0.00
G4_LPD	-0.04	0.02	-0.17	0.10	-1.62	0.10

	Step +in/-out	Multiple R	Multiple R-square	R-square change	F - to entr/rem	p-level	Variables included
L_BAKERSF	1	0.56	0.32	0.32	1064.96	0.00	1
G4_HPD	2	0.63	0.39	0.08	284.38	0.00	2
G2_7DMIN	3	0.66	0.43	0.04	172.82	0.00	3
G4_HPC	4	0.67	0.45	0.01	54.92	0.00	4
G2_1DMAX	5	0.67	0.45	0.00	10.55	0.00	5
G4_LPC	6	0.67	0.45	0.00	8.01	0.00	6
G4_LPD	7	0.67	0.45	0.00	2.63	0.10	7

	Beta	Std.Err. of Beta	B	Std.Err. of B	t(2043)	p-level
Intercept			15.90832	0.363814	43.72647	0.000000
AS_R_High(med)_res	0.650109	0.012786	0.47116	0.009266	50.84681	0.000000
AS_R_Commercial/Industrial/Transp	0.340614	0.012878	0.28683	0.010845	26.44904	0.000000
AS_R_Low Intensity Residential	0.143669	0.011542	0.11817	0.009493	12.44770	0.000000

	Step +in/-out	Multiple R	Multiple R-square	R-square change	F - to entr/rem	p-level	Variables included
AS_R_High(med)_res	1	0.793587	0.629780	0.629780	3478.737	0.000000	1
AS_R_Commercial/Industrial/Transp	2	0.844076	0.712464	0.082684	587.772	0.000000	2
AS_R_Low Intensity Residential	3	0.855999	0.732734	0.020270	154.945	0.000000	3

Table A-3. Correlation Coefficients (*r* Values) of Candidate Gradients Versus Primary Urban Gradient (Scaled and Averaged population Density, Road Density and %Urban LULC). All correlations significant ($P < 0.05$). Correlations were used for guiding determination or final alternative urban gradients.

	Urban Benchmark
Gradient 1	0.47
Gradient 2	0.59
Gradient 3	0.56
Gradient 4	0.44
Gradient 5	0.39
Gradient 6	0.83
Gradient 7	0.80
Gradient 8	0.76
Gradient 9	0.77
Gradient 10	0.79
Gradient 11	0.78
Gradient 12	0.66
Gradient 13	0.83
Gradient 14	0.56
Gradient 15	0.76
Gradient 16	0.83
Gradient 17	0.76

APPENDIX B

BIOLOGICAL METRICS CONSIDERED

Metric abbreviation	Metric	Type
SprwlTaxR100	Sprawler Richness	richness
Evenness	Evenness	
TanytTaxR100	Tanytarsini Richness	richness
Shan_10	Shannon Diversity	diversity
OrthoTaxR100	Orthocladiinae Richness	richness
Shan_e	Shannon Diversity	diversity
Shan_2	Shannon Diversity	diversity
CrMolTaxR100	Crustacean and Mollusk Richness	richness
SwmmrTaxR100	Swimmer Richness	richness
ColeoTaxR100	Coleoptera Richness	richness
TrichTaxR100	Trichoptera Richness	richness
PlecoTaxR100	Plecoptera Richness	richness
OligoTaxR100	Oligochaeta Richness	richness
HBI	Hilsenhoff Biotic Index	tolerance and biotic index
HBI_Urb	Hilsenhoff Biotic Index with Urban Tolerance	tolerance and biotic index
ClmbrTaxR100	Climber Richness	richness
FiltrTaxR100	Filterer Richness	richness
ShredTaxR100	Shreddder Richness	richness
D_Mg	Margalef Diversity	diversity
ScrapTaxR100	Scraper Richness	richness
EphemTax	Number of Ephemeroptera Taxa	richness
BrrwrTaxR100	Burrower Richness	richness
IntolTaxR100	Intolerant taxa richness	richness
TolerTaxR100	Tolerant taxa richness	richness
PredTaxR100	Predator richness	richness
ClngrTaxR100	Clinger richness	richness
EPTTaxR100	EPT richness	richness
IntolTaxR100Urb	Intolerant taxa richness with Urban Tolerance	tolerance and biotic index
CllctTaxR100	Collector richness	functional feeding group
TolerTaxR100Urb	Tolerant Taxa Richness with Urban Tolerance	tolerance and biotic index
BeckBIR100	Beck's Biotic Index	tolerance and biotic index
ChiroTaxR100	Chironomidae richness	richness
BeckBIR100Urb	Beck's Biotic Index with Urban Tolerance	tolerance and biotic index
DipTax	Number of Diptera taxa	richness
TotalTaxR100	Total number of taxa	richness
TanytPct	Percent Tanytarsini	compositional
OdonPct	Percent Odonata	compositional
GastrPct	Percent Gastropoda	compositional
ClmbrPct	Percent climber	compositional
SwmmrPct	Percent swimmer	compositional
IsoPct	Percent Isopoda	compositional
AmphPct	Percent Amphipoda	compositional
Baet2EphPct	Percent Baetidae/Ephemeroptera	compositional
BivalPct	Percent bivalvia	compositional
BrrwrPct	Percent burrower	compositional
ChiroPct	Percent Chironomidae	compositional
CllctPct	Percent collector	compositional
ClngrPct	Percent clinger	compositional
ColeoPct	Percent Coleoptera	compositional
CorbPct	Percent Corbicula	compositional
CrCh2ChiPct	Percent Chironomidae as Cricotopus and Chironomus	compositional
CrMolPct	Percent Crustacea and Mollusca	compositional
DipPct	Percent Diptera	compositional
Dom01Pct	Percent of Dominant Taxon	compositional
EphemPct	Percent Ephemeroptera	compositional
EPTPct	Percent EPT	compositional
FiltrPct	Percent filterer	compositional
Hyd2EPTPct	Percent of EPT as Hydropsychidae	compositional
Hyd2TriPct	Percent Trichoptera as Hydropsychidae	compositional
IntolPct	Percent intolerant	compositional
IntolPctUrb	Percent intolerant with Urban Tolerance	compositional
NonInPct	Percent non-insect	compositional
OligoPct	Percent Oligocheata	compositional
Orth2ChiPct	Percent Chironomidae as Orthocladiinae	compositional
PlecoPct	Percent Plecoptera	compositional
PredPct	Percent predator	compositional
ScrapPct	Percent scraper	compositional
ShredPct	Percent shredder	compositional
SprwlPct	Percent sprawler	compositional
Tnyt2ChiPct	Percent Chironomidae as Tanytarsini	compositional
TolerPct	Percent tolerant	compositional
TolerPctUrb	Percent tolerant with Urban Tolerance	compositional
TrichPct	Percent Trichoptera	compositional

IWA Publishing's authorised EU representative for General Product Safety Regulations is Diane D'Arras, 15 rue Duret, 75116 Paris, France, e-mail: safety@iwap.co.uk.

Printed and bound by CPI Group (UK) Ltd, Croydon, CR0 4YY
23/09/2026
02221511-0001